Vector Analysis
In Chemistry

McGRAW-HILL SERIES IN ADVANCED CHEMISTRY

BALLHAUSEN Introduction to Ligand Field Theory
BENSON The Foundations of Chemical Kinetics
DAVIDSON Statistical Mechanics
DEAN Flame Photometry
DEWAR The Molecular Orbital Theory of Organic Chemistry
ELIEL Stereochemistry of Carbon Compounds
FITTS Nonequilibrium Thermodynamics
FITTS Vector Analysis in Chemistry
HAMMETT Physical Organic Chemistry
HELFFERICH Ion Exchange
HINE Physical Organic Chemistry
JENCKS Catalysis in Chemistry and Enzymology
KAN Organic Photochemistry
LAIDLER Theories of Chemical Reaction Rates
LAITINEN Chemical Analysis
McDOWELL Mass Spectrometry
MANDELKERN Crystallization of Polymers
MARCH Advanced Organic Chemistry: Reactions, Mechanisms, and Structure
PITZER AND BREWER (*Revision of* Lewis and Randall) Thermodynamics
PLOWMAN Enzyme Kinetics
POPLE, SCHNEIDER, AND BERNSTEIN High-Resolution Nuclear Magnetic Resonance
POPLE AND BEVERIDGE Approximate Molecular Orbital Theory
PRYOR Free Radicals
RAAEN, ROPP, AND RAAEN Carbon-14
SHRIVER The Manipulation of Air-sensitive Compounds
SIGGIA Survey of Analytical Chemistry
WERTZ AND BOLTON Eletron Spin Resonance: Elementary Theory and Practical Applications
WIBERG Laboratory Technique in Organic Chemistry

Vector Analysis In Chemistry

Donald D. Fitts
Professor of Chemistry
University of Pennsylvania

McGraw-Hill Book Company
New York St. Louis San Francisco Düsseldorf Johannesburg Kuala Lumpur London Mexico Montreal New Delhi Panama Paris São Paulo Singapore Sydney Tokyo Toronto

This book was set in Press Roman by Publications Development Corporation.
The editors were Robert H. Summersgill, Joan Stern, and Janet Wagner;
the designer was Barbara Ellwood;
and the production supervisor was Thomas J. LoPinto.
The drawings were done by John Cordes, J & R Technical Services, Inc.
The Book Press, Inc., was printer and binder.

Library of Congress Cataloging in Publication Data

Fitts, Donald D
Vector analysis in chemistry.

(McGraw-Hill series in advanced chemistry)
1. Chemistry–Mathematics. 2. Vector analysis.
I. Title.
QD39.3.M3F57 540'.1'51563 73-16294
ISBN 0-07-021130-2

Vector Analysis
In Chemistry

1 2 3 4 5 6 7 8 9 0 B P B P 7 9 8 7 6 5 4

**To Beverly,
Robbie,
and Billy**

Contents

Preface

Vectors have permeated almost all areas of chemistry. Many physical phenomena of interest to chemists require the use of vector algebra and vector calculus in their mathematical representations. Accordingly, the purpose of this book is to introduce the concepts and methods of vector analysis and to illustrate their applications to areas of interest to chemists. This book is intended to serve both as a supplement or companion for physical chemistry, structural chemistry, and theoretical chemistry courses and as a text for independent study to be used throughout and beyond the chemistry student's academic career.

The basic mathematical topics covered here are vector algebra, vector calculus, dyadics, and vectors in orthogonal coordinate systems other than cartesian. Since spherical coordinates are often useful in chemical problems, a separate chapter is devoted to vector analysis in this coordinate system. It is hoped that this chapter on spherical coordinates will make the subsequent chapter on generalized orthogonal curvilinear coordinates seem less abstract.

The three areas selected to illustrate the applications of vector analysis are molecular geometry, electromagnetic theory, and fluid mechanics and thermodynamics. The application of vector algebra to molecular geometry shows how vectors may be used to facilitate the calculation of bond angles and atom-atom distances. The chapter on electromagnetic theory is deductive in approach, starting with Maxwell's equations and deducing the physical consequences of those postulates. The electromagnetic plane wave, linear and circular polarization, and optical activity are featured. Although most of the older and some of the current chemical literature uses the CGS system of units for electromagnetic phenomena, SI units are becoming more prevalent; accordingly, the equations in that chapter are developed in both sets of units. The chapter on fluid mechanics and thermodynamics develops the vector equations for the transport of mass, momentum, and energy in fluid systems. Also included in that chapter are discussions on the second law of thermodynamics for open systems and on sound waves in fluids.

Each chapter closes with a short selection of problems on which the reader can test his understanding of the material in the chapter. For easy reference a summary of various vector and dyadic relationships is given in the Appendix.

Donald D. Fitts

Vector Analysis in Chemistry

Chapter one

Vector Algebra

1-1 Introduction

Since the world in which we live has three space dimensions, the fundamental physical laws are mathematical relationships involving these three dimensions. In order to express and analyze these relationships in a convenient and compact manner, the branch of mathematics known as vector analysis was devised. The purpose of this book is to develop the basic definitions, operations, and theorems of vector analysis so that the concepts and procedures may be applied to problems of chemical interest.

A *scalar* is a quantity which has only a magnitude, whereas a *vector* is a quantity with both a magnitude and a direction. The mass of a body and its uniform temperature are examples of a scalar. The velocity of a body and the force acting upon it are vector quantities. A scalar may be positive or negative, real or complex, but the magnitude of a vector is a positive, real number.

A *scalar function* or *field* is a scalar quantity associated with each point in a region of space. For example, the temperature distribution in a stream of water is a scalar function. A *vector function* or *field* is a vector quantity associated with each point in a region of space. An example of a vector function is the velocity distribution of water in a stream. The water in the center moves faster than the water at the edges; that near the surface faster than that which is deeper. Furthermore, the directions of the vectors at each of the points need not be the same. Thus, in this example, eddy currents are the result of local velocities which are not parallel to the main direction of flow of the stream of water.

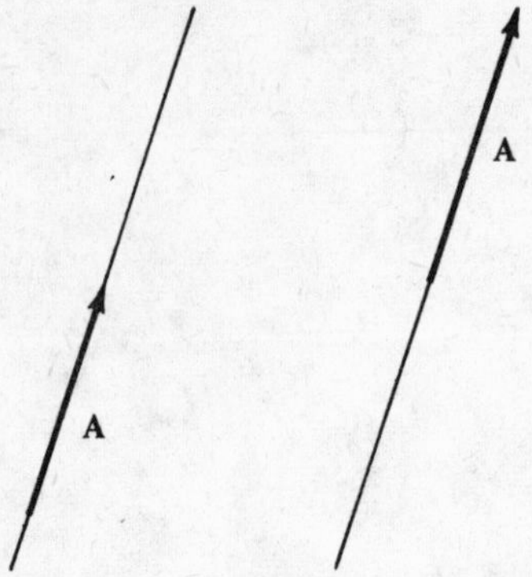

FIGURE 1-1

1-2 Representation of a vector

A vector can be represented as a directed line segment in space. Two vectors are equal if their magnitudes and directions are the same. Thus, parallel directed line segments of the same length and direction may represent the same vector (see Fig. 1-1). Vectors and vector fields are usually denoted by boldface type (**A**). In handwriting, a vector quantity may be denoted by an arrow over the symbol ($\vec{A}$).

A vector can be resolved into its three components in the directions of the axes of a cartesian coordinate system. Thus, the vector **A** has a projection A_x on the x axis, A_y on the y axis, and A_z on the z axis, as shown in Fig. 1-2. From the pythagorean theorem, the magnitude A of the vector **A** is given by

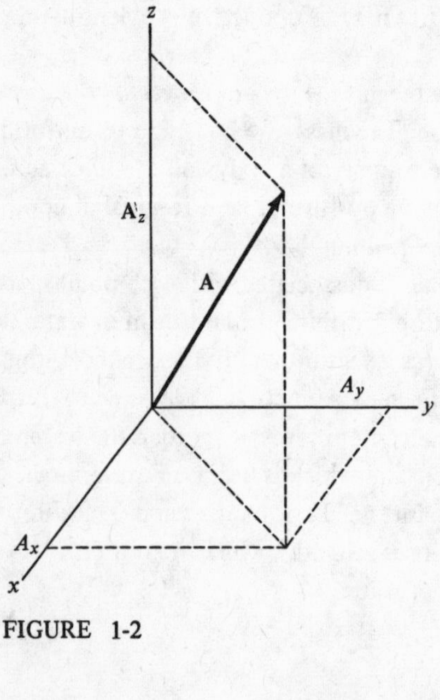

FIGURE 1-2

$$A = \sqrt{A_x^2 + A_y^2 + A_z^2} \tag{1-1}$$

We next introduce a set of *unit vectors* **i**, **j**, **k** to denote the positive directions of the x, y, and z axes, respectively. Thus, **i** is a vector of magnitude *one*, directed along the positive x axis, as illustrated in Fig. 1-3. The unit vectors **j** and **k** are similarly directed. In terms of these unit vectors, the vector **A** may be represented as

$$\mathbf{A} = \mathbf{i}A_x + \mathbf{j}A_y + \mathbf{k}A_z \tag{1-2}$$

Two vectors are equal if their corresponding components are equal. A vector is zero or vanishes if its magnitude is zero, i.e., if all three of its components are zero.

Throughout this book only *right-handed* cartesian coordinate systems are used. If (**i**, **j**, **k**) denotes an ordered set of mutually orthogonal (perpendicular) unit vectors, then *right-handed* is defined in absolute terms by letting the thumb, index finger, and middle finger of the right hand represent, respectively, the first, second, and third elements of the ordered set. Thus, Fig. 1-3 illustrates a right-handed cartesian coordinate system.

Let (**i**, **j**, **k**) denote a fixed set of unit vectors, corresponding to a fixed right-handed xyz coordinate system. Let (**i**′, **j**′, **k**′) denote any other set of mutually perpendicular vectors in space. If the original set (**i**, **j**, **k**) of unit vectors can be moved rigidly through space and made to coincide with the set (**i**′, **j**′, **k**′) so that **i** coincides with **i**′, **j** with **j**′, and **k** with **k**′, then (**i**′, **j**′, **k**′) can also serve as the basis of a *right-handed* coordinate system. If, on the other hand, **i** coincides with **i**′, and **j** with **j**′, but **k** coincides with −**k**′, then the set (**i**′, **j**′, **k**′) can serve as the basis of a *left-handed* coordinate system. From this consideration it follows that the ordered sets (**i**, **j**, **k**), (**j**, **k**, **i**), and (**k**, **i**, **j**) lead to right-handed coordinate

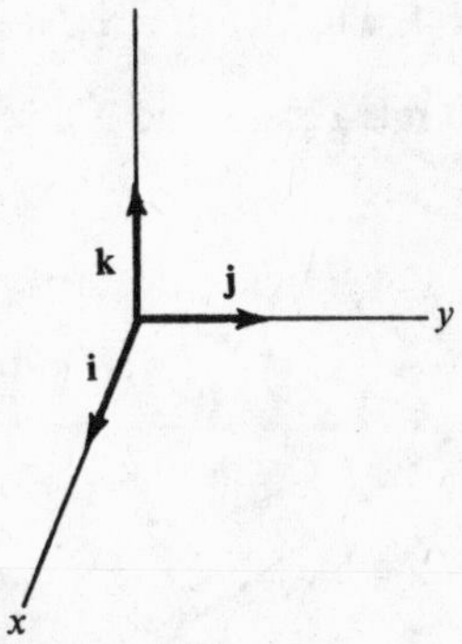

FIGURE 1-3

systems, and the ordered sets (**i**, **k**, **j**), (**j**, **i**, **k**), and (**k**, **j**, **i**) yield left-handed coordinate systems.

1-3 Addition and subtraction of vectors

Consider two vectors **A** and **B** given by

$$\begin{aligned} \mathbf{A} &= \mathbf{i}A_x + \mathbf{j}A_y + \mathbf{k}A_z \\ \mathbf{B} &= \mathbf{i}B_x + \mathbf{j}B_y + \mathbf{k}B_z \end{aligned} \tag{1-3}$$

The sum and difference of these two vectors are

$$\begin{aligned} \mathbf{A}+\mathbf{B} &= \mathbf{i}(A_x + B_x) + \mathbf{j}(A_y + B_y) + \mathbf{k}(A_z + B_z) \\ \mathbf{A}-\mathbf{B} &= \mathbf{i}(A_x - B_x) + \mathbf{j}(A_y - B_y) + \mathbf{k}(A_z - B_z) \end{aligned} \tag{1-4}$$

Thus, the sum (difference) of two vectors is obtained by adding (subtracting) the corresponding components.

The geometrical significance of the sum and difference of two vectors is illustrated in Figs. 1-4 and 1-5. Note that the subtraction of **B** from **A** is equivalent to the addition of **A** and −**B**, where −**B** is a vector with the same magnitude as **B** but of opposite direction. The three components of −**B** have the opposite sign to the corresponding components of **B**.

From the definitions (1-4) we see that vector addition obeys the commutative law,

$$\mathbf{A}+\mathbf{B} = \mathbf{B}+\mathbf{A} \tag{1-5}$$

and the associative law,

$$(\mathbf{A}+\mathbf{B})+\mathbf{C} = \mathbf{A}+(\mathbf{B}+\mathbf{C}) \tag{1-6}$$

1-4 Product of a scalar and a vector

If α is a scalar and **A** a vector, the product $\alpha\mathbf{A}$ is defined as

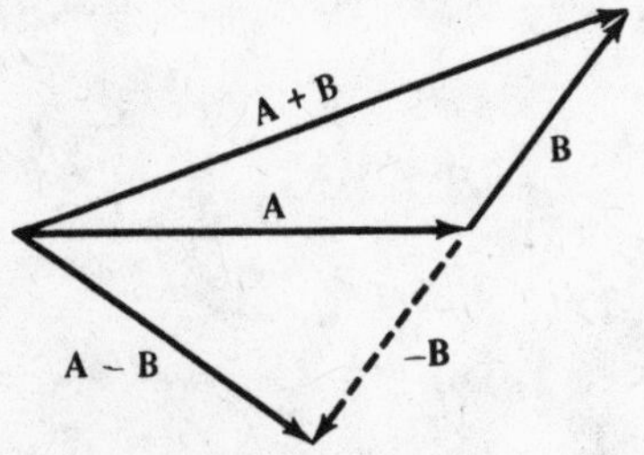

FIGURE 1-4

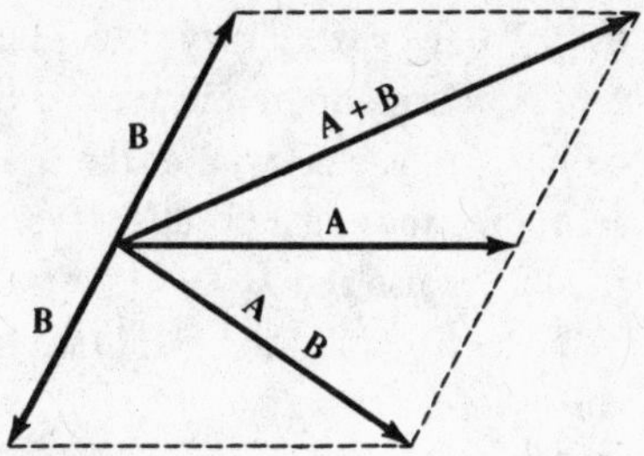

FIGURE 1-5

$$\alpha\mathbf{A} = \mathbf{i}\alpha A_x + \mathbf{j}\alpha A_y + \mathbf{k}\alpha A_z \tag{1-7}$$

If α is positive, the new vector $\alpha\mathbf{A}$ is in the same direction as **A**. If α is negative, $\alpha\mathbf{A}$ is in the direction of $-\mathbf{A}$. If α is zero, then $\alpha\mathbf{A}$ is zero.

Two vectors **A** and **B** are *collinear* or *linearly dependent* if scalars α and β, not zero, exist such that

$$\alpha\mathbf{A} + \beta\mathbf{B} = 0 \tag{1-8}$$

This expression is equivalent to the equation

$$\mathbf{A} = a\mathbf{B} \tag{1-9}$$

and states that **A** and **B** are represented by parallel line segments. Three vectors **A**, **B**, and **C** are *coplanar* or *linearly dependent* if there exist scalars α, β, and γ, not zero, such that

$$\alpha\mathbf{A} + \beta\mathbf{B} + \gamma\mathbf{C} = 0 \tag{1-10}$$

1-5 Scalar product of two vectors

The *scalar product* or *dot product* of two vectors **A** and **B** is a scalar quantity denoted by $\mathbf{A} \cdot \mathbf{B}$ and defined as

$$\mathbf{A} \cdot \mathbf{B} = AB \cos\theta \tag{1-11}$$

where θ is the angle between the vectors **A** and **B**. Since $\cos\theta$ is equal to

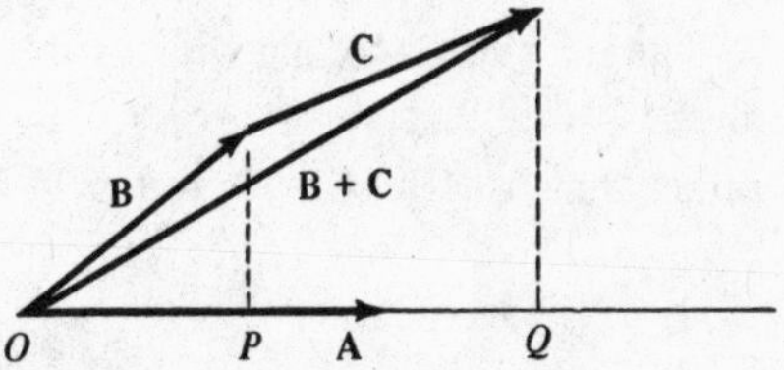

FIGURE 1-6

cos $(-\theta)$, the angle θ may be considered the angle *from* **A** *to* **B** or the angle *from* **B** *to* **A**. Furthermore, since cos θ is equal to cos $(2\pi - \theta)$, the angle θ may be defined in the clockwise or the counterclockwise direction. If either **A** or **B** is zero, then the scalar product is zero. If **A** is perpendicular to **B**, the scalar product is also zero.

From the definition (1-11) we see that the scalar product obeys the *commutative law:*

$$\mathbf{A} \cdot \mathbf{B} = \mathbf{B} \cdot \mathbf{A} \tag{1-12}$$

That the scalar product also obeys the *distributive law,*

$$\mathbf{A} \cdot (\mathbf{B} + \mathbf{C}) = \mathbf{A} \cdot \mathbf{B} + \mathbf{A} \cdot \mathbf{C} \tag{1-13}$$

may be seen with the aid of Fig. 1-6. We have from the figure

$$\begin{aligned} \mathbf{A} \cdot \mathbf{B} + \mathbf{A} \cdot \mathbf{C} &= A(OP) + A(PQ) \\ &= A(OQ) \\ &= \mathbf{A} \cdot (\mathbf{B} + \mathbf{C}) \end{aligned} \tag{1-14}$$

where the line segment (OP) is the projection of **B** onto **A**, (PQ) the projection of **C** onto **A**, and (OQ) the projection of **B** + **C** onto **A**.

We next apply the definition of the scalar product to the set of unit vectors **i, j, k** and obtain the following results:

$$\begin{array}{lll} \mathbf{i} \cdot \mathbf{i} = 1 & \mathbf{j} \cdot \mathbf{j} = 1 & \mathbf{k} \cdot \mathbf{k} = 1 \\ \mathbf{i} \cdot \mathbf{j} = 0 & \mathbf{i} \cdot \mathbf{k} = 0 & \mathbf{j} \cdot \mathbf{k} = 0 \end{array} \tag{1-15}$$

If we write **A** and **B** in terms of their components as in Eqs. (1-3), we may expand the scalar product, using the distributive law (1-13) and the relations (1-15) to obtain

$$\mathbf{A} \cdot \mathbf{B} = A_x B_x + A_y B_y + A_z B_z \tag{1-16}$$

By setting **B** = **A** in Eq. (1-16), we obtain

$$\mathbf{A} \cdot \mathbf{A} = A_x^2 + A_y^2 + A_z^2 = A^2 \tag{1-17}$$

Thus, the scalar product of **A** with itself gives the square of the magnitude of **A**.

The rules of vector algebra are not always the same as the rules of scalar algebra. Note that the relation **A** · **B** = **A** · **C** does *not* mean that **B** = **C** but rather that **A** · (**B** − **C**) = 0, so that **A** is perpendicular to the vector **B** − **C**.

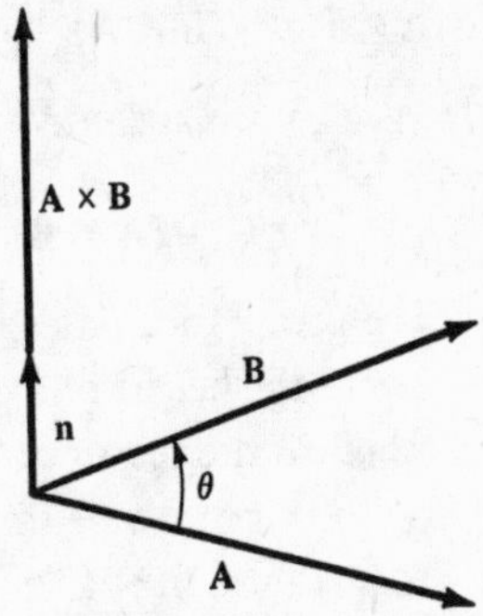

FIGURE 1-7

1-6 Vector product of two vectors

The *vector product* or *cross product* of two vectors **A** and **B**, written as **A** × **B**, is defined as a vector whose magnitude is the product of the magnitudes A and B and the sine of the angle between **A** and **B** and whose direction is perpendicular to the plane determined by **A** and **B**; thus,

$$\mathbf{A} \times \mathbf{B} = AB \sin \theta \, \mathbf{n} \tag{1-18}$$

where **n** is a unit vector normal to the plane of **A** and **B**. The direction of **n** is obtained by the right-hand rule: If the fingers of the right hand are pointed along the vector **A** and then closed through the angle θ toward the vector **B**, the thumb points in the positive direction of **n** (see Fig. 1-7).

The angle between **A** and **B** may be defined in either the clockwise or counterclockwise sense. To illustrate this fact, consider Fig. 1-8, where the angle between **A** and **B** is defined in the opposite sense to that in Fig. 1-7. Thus, the vector product is

$$\mathbf{A} \times \mathbf{B} = AB \sin \theta' \, \mathbf{n}' \tag{1-19}$$

However, we have

$$\theta' = 2\pi - \theta \tag{1-20a}$$

$$\sin \theta' = \sin (2\pi - \theta) = - \sin \theta \tag{1-20b}$$

$$\mathbf{n}' = - \mathbf{n} \tag{1-20c}$$

Consequently, substitution of Eqs. (1-20*b*) and (1-20*c*) into Eq. (1-19) yields

$$\mathbf{A} \times \mathbf{B} = AB \sin \theta \, \mathbf{n} \tag{1-21}$$

in agreement with Fig. 1-7.

Since $\sin \theta$ equals $-\sin(-\theta)$, the vector product obeys an *anticommutative* law:

$$\mathbf{A} \times \mathbf{B} = -(\mathbf{B} \times \mathbf{A}) \tag{1-22}$$

However, the *distributive law* is obeyed:

$$\mathbf{A} \times (\mathbf{B} + \mathbf{C}) = (\mathbf{A} \times \mathbf{B}) + (\mathbf{A} \times \mathbf{C}) \tag{1-23}$$

The geometrical interpretation of the vector product is straightforward. Since $B \sin \theta$ is the height of a parallelogram with adjacent sides **A** and **B**, the magnitude $AB \sin \theta$ of the vector product equals the area of this parallelogram (see Fig. 1-9). It follows from the definition (1-18) that the vector product of two parallel vectors is zero. Furthermore, the vector product of a vector with itself vanishes:

$$\mathbf{A} \times \mathbf{A} = 0 \tag{1-24}$$

The various vector products formed with the unit vectors **i, j, k** are

$$\begin{array}{lll} \mathbf{i} \times \mathbf{i} = 0 & \mathbf{j} \times \mathbf{j} = 0 & \mathbf{k} \times \mathbf{k} = 0 \\ \mathbf{i} \times \mathbf{j} = \mathbf{k} & \mathbf{j} \times \mathbf{k} = \mathbf{i} & \mathbf{k} \times \mathbf{i} = \mathbf{j} \end{array} \tag{1-25}$$

If we write the vectors **A** and **B** in terms of their components as in Eqs. (1-3) and apply the distributive law (1-23), we obtain

$$\mathbf{A} \times \mathbf{B} = \mathbf{i}(A_yB_z - A_zB_y) + \mathbf{j}(A_zB_x - A_xB_z) + \mathbf{k}(A_xB_y - A_yB_x) \tag{1-26}$$

This result is also given by the determinant

$$\mathbf{A} \times \mathbf{B} = \begin{vmatrix} \mathbf{i} & \mathbf{j} & \mathbf{k} \\ A_x & A_y & A_z \\ B_x & B_y & B_z \end{vmatrix} \tag{1-27}$$

1-7 Triple scalar product

The *triple scalar product* is

$$\mathbf{A} \cdot \mathbf{B} \times \mathbf{C}$$

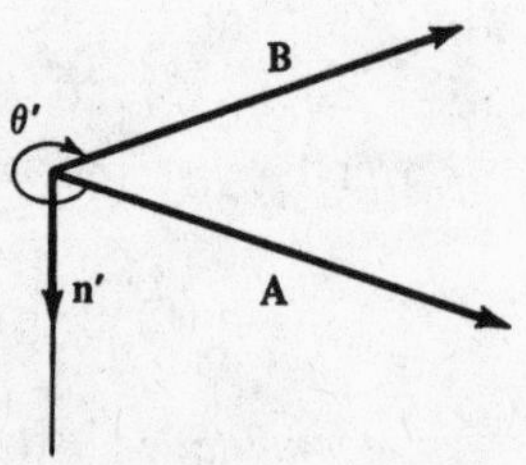

FIGURE 1-8

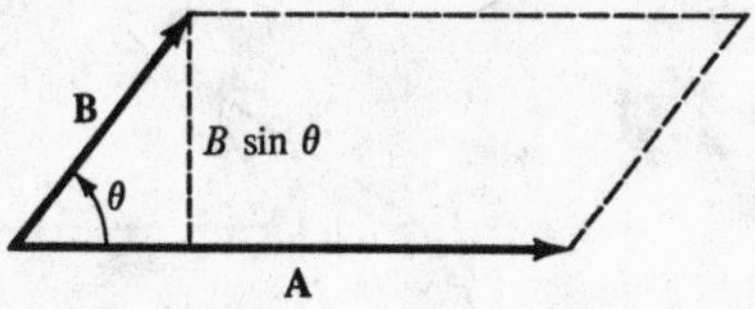

FIGURE 1-9

Parentheses here are not necessary because the expression $(\mathbf{A} \cdot \mathbf{B}) \times \mathbf{C}$ has no meaning. Since the scalar product obeys the commutative law (1-12), we also have

$$\mathbf{A} \cdot \mathbf{B} \times \mathbf{C} = \mathbf{B} \times \mathbf{C} \cdot \mathbf{A} \tag{1-28}$$

The vector product $\mathbf{B} \times \mathbf{C}$ is a vector normal to the plane of **B** and **C** and has a magnitude equal to the area of the parallelogram with **B** and **C** as adjacent sides. If **A** is not in the plane of **B** and **C** and forms an acute angle with $(\mathbf{B} \times \mathbf{C})$, then the scalar product of **A** and $(\mathbf{B} \times \mathbf{C})$ is simply the volume of the parallelepiped with edges **A**, **B**, **C**. Thus, we have

$$\mathbf{A} \cdot \mathbf{B} \times \mathbf{C} = ABC \cos\theta \ \sin\phi \tag{1-29}$$

where the angles θ and ϕ are defined in Fig. 1-10. If the angle between **A** and $(\mathbf{B} \times \mathbf{C})$ is obtuse, then $\mathbf{A} \cdot \mathbf{B} \times \mathbf{C}$ is the negative of the volume of the parallelepiped. It follows from Fig. 1-10 that

$$\mathbf{A} \cdot \mathbf{B} \times \mathbf{C} = \mathbf{B} \cdot \mathbf{C} \times \mathbf{A} = \mathbf{C} \cdot \mathbf{A} \times \mathbf{B} \tag{1-30}$$

since each of these products gives the same volume. We see from Eq. (1-29) that the triple scalar product vanishes when **A**, **B**, and **C** are coplanar, for then θ is 90°.

The triple scalar product can also be written as a determinant as follows:

$$\mathbf{A} \cdot \mathbf{B} \times \mathbf{C} = \mathbf{A} \cdot \begin{vmatrix} \mathbf{i} & \mathbf{j} & \mathbf{k} \\ B_x & B_y & B_z \\ C_x & C_y & C_z \end{vmatrix}$$

$$= A_x \begin{vmatrix} B_y & B_z \\ C_y & C_z \end{vmatrix} - A_y \begin{vmatrix} B_x & B_z \\ C_x & C_z \end{vmatrix} + A_z \begin{vmatrix} B_x & B_y \\ C_x & C_y \end{vmatrix}$$

$$= \begin{vmatrix} A_x & A_y & A_z \\ B_x & B_y & B_z \\ C_x & C_y & C_z \end{vmatrix} \tag{1-31}$$

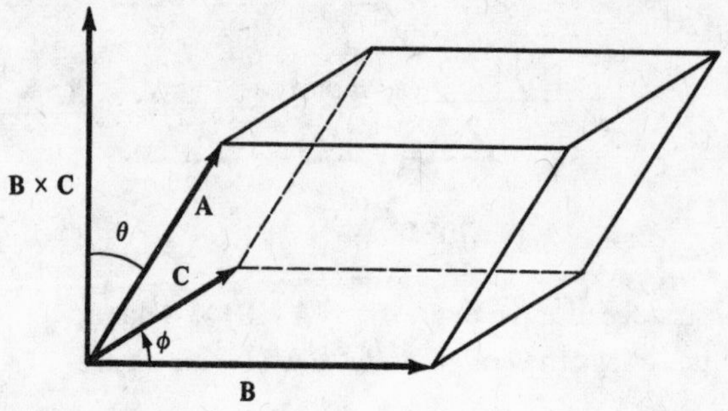

FIGURE 1-10

1-8 Triple vector product

The expressions

$$\mathbf{A} \times (\mathbf{B} \times \mathbf{C}) \qquad \text{and} \qquad (\mathbf{A} \times \mathbf{B}) \times \mathbf{C}$$

are called triple vector products. The parentheses are necessary in this case since the two expressions are not equal. The product $\mathbf{A} \times (\mathbf{B} \times \mathbf{C})$ is a vector in the plane of **B** and **C**, whereas the product $(\mathbf{A} \times \mathbf{B}) \times \mathbf{C}$ is a vector in the plane of **A** and **B**.

That $\mathbf{A} \times (\mathbf{B} \times \mathbf{C})$ lies in the plane of **B** and **C** can be easily seen. The vector $\mathbf{B} \times \mathbf{C}$ is perpendicular to both **B** and **C** and hence to the plane determined by **B** and **C**. (See Fig. 1-10.) The cross product of **A** and $(\mathbf{B} \times \mathbf{C})$ is perpendicular to $(\mathbf{B} \times \mathbf{C})$ (also to **A**) and hence must lie in the plane of **B** and **C**.

Since $\mathbf{A} \times (\mathbf{B} \times \mathbf{C})$ lies in the plane determined by **B** and **C**, this vector can be written as a linear combination of **B** and **C**:

$$\mathbf{A} \times (\mathbf{B} \times \mathbf{C}) = \alpha\mathbf{B} + \beta\mathbf{C} \tag{1-32}$$

The coefficients can be determined through expansion of the triple vector product in terms of the vector components. The result is that $\alpha = \mathbf{A} \cdot \mathbf{C}$ and $\beta = -\mathbf{A} \cdot \mathbf{B}$, so that

$$\mathbf{A} \times (\mathbf{B} \times \mathbf{C}) = (\mathbf{A} \cdot \mathbf{C})\mathbf{B} - (\mathbf{A} \cdot \mathbf{B})\mathbf{C} \tag{1-33}$$

PROBLEMS

1. An airplane has an air speed of 250 mi/h and is flying a course (relative to the ground) due north. Determine the speed of the airplane relative to the ground and the direction the pilot must fly in order to stay on course if the wind velocity is 50 mi/h
   (*a*) Coming *from* the west.
   (*b*) Coming *from* a northeasterly direction at a 30° angle from north.

2. Using vector addition and subtraction, show that the diagonals of a parallelogram bisect each other. (*Hint:* From an arbitrary origin construct two vectors, one to the midpoint of each diagonal, and then show that these two vectors are equal. Alternatively, with the aid of Fig. 1-5, set the two vectors drawn from the point of intersection of the diagonals to two adjacent vertices proportional to the corresponding diagonals and then show that the two proportionality constants are each equal to one-half.)

3. Find the angle between the x axis and the vector
$\mathbf{A} = 3\mathbf{i} + 2\mathbf{j} - 6\mathbf{k}$

4. The vectors **A** and **B** are given by
$\mathbf{A} = 2\mathbf{i} + 3\mathbf{j} + 7\mathbf{k}$
$\mathbf{B} = -\mathbf{i} + 6\mathbf{j} - 5\mathbf{k}$
Compute the scalar and vector products of **A** and **B**. Find the angle between **A** and **B**.

5. Show that the three vectors
$\mathbf{A} = 3\mathbf{i} + 2\mathbf{j} + 4\mathbf{k}$
$\mathbf{B} = -2\mathbf{i} + 5\mathbf{j}$
$\mathbf{C} = -\mathbf{i} - 7\mathbf{j} - 4\mathbf{k}$
when placed successively form a closed triangle. Calculate the three interior angles.

6. The vectors **A** and **B** are given by
$\mathbf{A} = \mathbf{i}\cos\theta + \mathbf{j}\sin\theta$
$\mathbf{B} = \mathbf{i}\cos\phi - \mathbf{j}\sin\phi$
and are shown in Fig. 1-11. Form the scalar and vector products of **A** and **B** and thereby derive the addition formulas for the sine and for the cosine.

7. Demonstrate that
$(\mathbf{A}\cdot\mathbf{B})^2 + |\mathbf{A}\times\mathbf{B}|^2 = (\mathbf{A}\cdot\mathbf{A})(\mathbf{B}\cdot\mathbf{B})$

8. Prove that
$(\mathbf{A}+\mathbf{B})\cdot(\mathbf{A}-\mathbf{B}) = A^2 - B^2$
and
$(\mathbf{A}+\mathbf{B})\times(\mathbf{A}-\mathbf{B}) = 2\mathbf{B}\times\mathbf{A}$

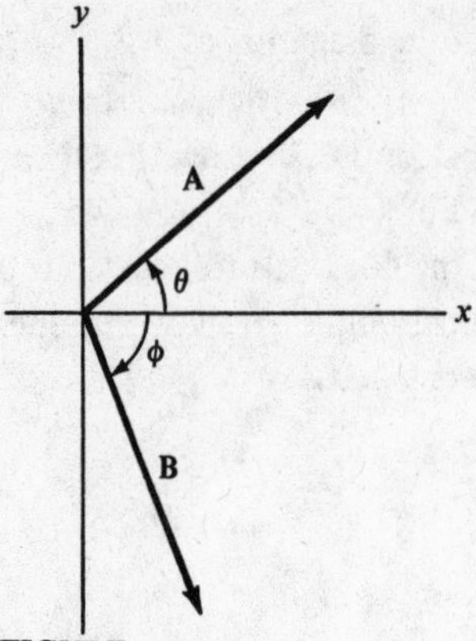

FIGURE 1-11

9. Three vectors **A**, **B**, and **C** are given by
   A = 2**i** − 3**j** + 2**k**
   B = 4**i** + 5**j** − 2**k**
   C = −2**i** − 3**j** − 5**k**
   Compute the values of **A** · **B** × **C** and **A** × (**B** × **C**).

10. Show that the coefficients α and β in Eq. (1-32) are those given in Eq. (1-33).

Chapter two

Molecular Geometry

Among the many practical uses of vector algebra is its application to the determination of molecular geometry. Although the required calculations in this area of application can be carried out by using only three-dimensional geometry and trigonometry, vector algebra simplifies many of the calculations and reduces the necessity for imaginative geometric visualization. In this chapter some illustrative examples of the application of vector algebra to molecular geometry are presented.

2-1 Bond angles and distances

The problem of determining bond angles and distances between atoms in a molecule may be readily solved by means of vector algebra. The first step in such a procedure is to establish a cartesian coordinate system. It is usually convenient to select the center of one of the atoms as origin and to let the x axis extend along one of the bonds connecting that atom to another. The y axis should be chosen so that as many atoms of the molecule as possible lie in the xy plane. The z axis is determined, of course, by the right-hand rule. Often it is best to make use of the elements of molecular symmetry in locating the cartesian coordinate axes. Thus, a coordinate axis may be constructed to lie in a plane of symmetry.

Once the coordinate system is selected, a vector is constructed along each bond in the molecule. The magnitude of the vector is the bond length. These vectors are best directed so that the head of one is the tail (origin) of the next

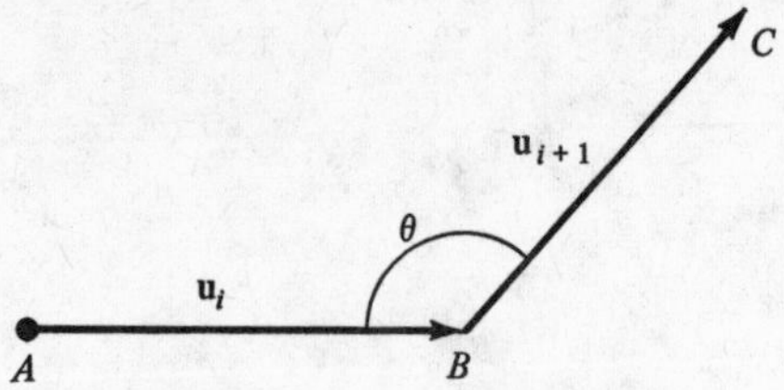

FIGURE 2-1

(see Fig. 2-1). Thus, in a cyclic molecule the sum of the vectors around the cycle is zero.

The major problem in the application of the vector algebra to molecular geometry is to determine the three components of these bond vectors. Here one must use a knowledge of bond lengths, bond angles, and molecular symmetry The first bond vector $\mathbf{u}_1$ may, for example, lie on the x axis so that it is $\mathbf{i}\,a_1$, where a_1 is the bond length. The second bond vector $\mathbf{u}_2$ has magnitude a_2, makes a bond angle θ with $\mathbf{u}_1$, and may lie in the xy plane. Thus, its components $\alpha_x, \alpha_y, \alpha_z$ are determined by

$$\alpha_x{}^2 + \alpha_y{}^2 + \alpha_z{}^2 = a_2{}^2 \tag{2-1a}$$

$$(-\mathbf{u}_1) \cdot \mathbf{u}_2 = a_1 a_2 \cos\theta \tag{2-1b}$$

$$\alpha_z = 0 \tag{2-1c}$$

The minus sign in Eq. (2-1b) arises because θ is the angle between $\mathbf{u}_2$ and $-\mathbf{u}_1$ as shown in Fig. 2-1. Equations (2-1) readily yield the result

$$\mathbf{u}_2 = -\mathbf{i}\,a_2 \cos\theta \pm \mathbf{j}\,a_2 \sin\theta \tag{2-2}$$

The sign of the y component depends on whether $\mathbf{u}_2$ points in the positive or negative y direction.

The components of the third bond vector $\mathbf{u}_3$ are more difficult to determine. The magnitude of $\mathbf{u}_3$ is a_3, the appropriate bond length. Moreover, $\mathbf{u}_3$ makes a presumably known bond angle θ with $\mathbf{u}_2$, so that

$$(-\mathbf{u}_2) \cdot \mathbf{u}_3 = a_2 a_3 \cos\theta \tag{2-3}$$

Now, however, the coordinate axes have all been chosen, so that we have only two relationships but three unknowns, namely, the three components of $\mathbf{u}_3$. Thus we must have additional geometrical information regarding the structure of the molecule in order to determine the remaining bond vectors. In succeeding sections of this chapter we consider specific examples and discuss the solutions to these determinations of bond vectors.

Once all the bond vectors in the molecule are determined, it is possible to calculate any distance or any angle by means of vector summation, the scalar product, or the cross product. These procedures are best explained by means of specific examples, which appear in the following sections.

2-2 Benzene

A particularly simple example is the benzene molecule, the carbon atoms of which describe a regular hexagon. The carbon-carbon bond lengths a are all the same and the interior angles are all 120°. We will not consider the hydrogen atoms at first but defer such a discussion to the end of this section. We label the carbon atoms $A, B, \ldots, F$ and select A as the origin of the cartesian coordinate system. The x axis is taken along the bond (AB), the y axis along the line (not a bond) (AE). Since the molecule is planar, we have no need for the z axis.

The bond vectors $\mathbf{u}_1, \mathbf{u}_2, \ldots, \mathbf{u}_6$ are constructed as shown in Fig. 2-2. Since $\mathbf{u}_1$ is along the positive x axis, we have

$$\mathbf{u}_1 = \mathbf{i}\,a \tag{2-4a}$$

The bond vector $\mathbf{u}_2$ is readily determined by trigonometric considerations:

$$\mathbf{u}_2 = \mathbf{i}\,a \cos 60° + \mathbf{j}\,a \sin 60° = \mathbf{i}\frac{a}{2} + \mathbf{j}\frac{\sqrt{3}}{2}\,a \tag{2-4b}$$

Since atom D has the same x coordinate as B and since the y coordinate of C is halfway between the y coordinates of B and D, the x component of $\mathbf{u}_3$ is the negative of the x component of $\mathbf{u}_2$ and the y components of $\mathbf{u}_2$ and $\mathbf{u}_3$ are the same, so that

$$\mathbf{u}_3 = -\mathbf{i}\frac{a}{2} + \mathbf{j}\frac{\sqrt{3}}{2}\,a \tag{2-4c}$$

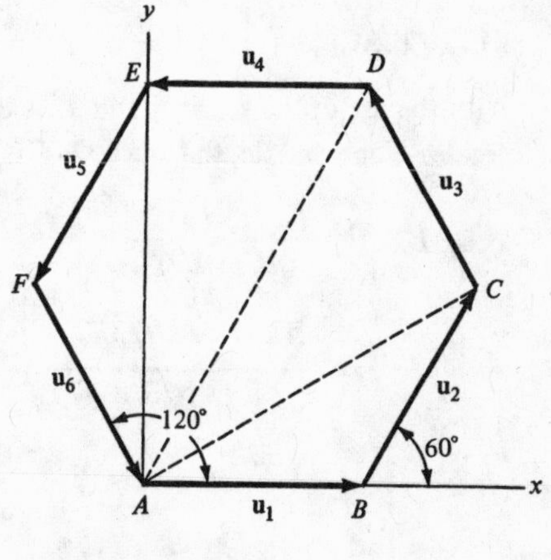

FIGURE 2-2

The vectors $\mathbf{u}_4, \mathbf{u}_5, \mathbf{u}_6$ are antiparallel, respectively, to $\mathbf{u}_1, \mathbf{u}_2, \mathbf{u}_3$:

$$\mathbf{u}_4 = -\,\mathbf{u}_1 = -\,\mathbf{i}\,a \tag{2-4d}$$

$$\mathbf{u}_5 = -\,\mathbf{u}_2 = -\,\mathbf{i}\frac{a}{2} - \mathbf{j}\frac{\sqrt{3}}{2}\,a \tag{2-4e}$$

$$\mathbf{u}_6 = -\,\mathbf{u}_3 = \mathbf{i}\frac{a}{2} - \mathbf{j}\frac{\sqrt{3}}{2}\,a \tag{2-4f}$$

Note that the sum of the vectors around the cycle *ABCDEFA* vanishes:

$$\mathbf{u}_1 + \mathbf{u}_2 + \mathbf{u}_3 + \mathbf{u}_4 + \mathbf{u}_5 + \mathbf{u}_6 = 0 \tag{2-5}$$

In this example the origin of the coordinate system was fixed at atom *A*. The resulting bond vectors as given in Eqs. (2-4) are, of course, independent of this choice of origin. We could just as well have chosen some other point, say atom *F*, as origin. The components of the set of vectors (2-4) do depend, however, on the choice of orientation for the *x* axis. A different orientation will yield a different, but equally valid, set of bond vectors.

From a knowledge of the bond vectors (2-4), we can determine any other geometric property of the benzene molecule. For example, we can determine the distance (*AD*). Let (**AD**) represent the vector drawn from *A* to *D*. Then we have

$$(\mathbf{AD}) = \mathbf{u}_1 + \mathbf{u}_2 + \mathbf{u}_3 = \mathbf{i}a + \mathbf{j}\sqrt{3}a$$

and

$$(AD) = |(\mathbf{AD})| = (a^2 + 3a^2)^{1/2} = 2a$$

The distance (*AC*) can also be readily determined:

$$(\mathbf{AC}) = \mathbf{u}_1 + \mathbf{u}_2 = \mathbf{i}\frac{3}{2}a + \mathbf{j}\frac{\sqrt{3}}{2}a$$

$$(AC) = |(\mathbf{AC})| = \left(\frac{9}{4}a^2 + \frac{3}{4}a^2\right)^{1/2} = \sqrt{3}\,a$$

With the use of the scalar product we can readily calculate angles in the molecule. For example, the angle *DAC* is determined by

$$\begin{aligned}\cos\angle DAC &= \frac{(\mathbf{AC})\cdot(\mathbf{AD})}{(AC)\,(AD)} \\ &= \frac{[\mathbf{i}\frac{3}{2}a + \mathbf{j}(\sqrt{3}/2)a]\cdot(\mathbf{i}a + \mathbf{j}\sqrt{3}\,a)}{(\sqrt{3}\,a)(2a)} \\ &= \sqrt{3}/2\end{aligned}$$

so that

$$\angle DAC = \cos^{-1}\frac{\sqrt{3}}{2} = 30°$$

Similarly, the angle ACD is determined by

$$\cos \angle ACD = \frac{(\mathbf{CA}) \cdot (\mathbf{CD})}{(CA)\,(CD)}$$

Since $(\mathbf{CD}) = \mathbf{u}_3$ and $(\mathbf{CA}) = -(\mathbf{AC})$, we find that

$$\cos \angle ACD = 0$$

or

$$\angle ACD = 90°$$

We now consider the hydrogen atoms on the benzene molecule. One hydrogen atom is bonded to each carbon atom so that any angle formed by carbon-carbon-hydrogen is 120°. The carbon-hydrogen bond lengths all have the same value b. We label the hydrogen atoms $A', B', \ldots, F'$, with A' bonded to carbon atom A, etc. (See Fig. 2-3.) Let $\mathbf{w}_A$ be the bond vector from carbon atom A to hydrogen atom A' and so forth for $\mathbf{w}_B, \ldots, \mathbf{w}_F$. We again note that the benzene molecule is planar, so that the six bond vectors $\mathbf{w}_i$ have no z components.

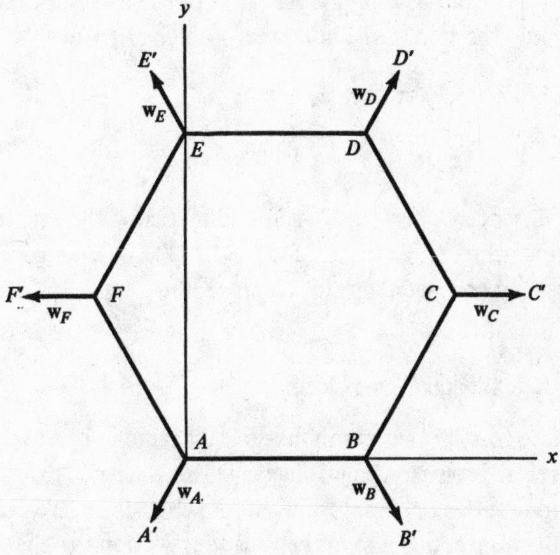

FIGURE 2-3

The vector $\mathbf{w}_A$ with components α_x and α_y ($\alpha_z = 0$) forms 120° angles with $\mathbf{u}_1$ and with $(-\mathbf{u}_6)$, and so we may write

$$\mathbf{u}_1 \cdot \mathbf{w}_A = a\alpha_x + 0 \cdot \alpha_y = ab \cos 120^\circ = -\frac{1}{2}ab$$

$$(-\mathbf{u}_6) \cdot \mathbf{w}_A = -\frac{1}{2}a\alpha_x + \frac{\sqrt{3}}{2}a\alpha_y = ab \cos 120^\circ = -\frac{1}{2}ab$$

from which it follows that $\alpha_x = -b/2$ and $\alpha_y = -\sqrt{3}\,b/2$ or

$$\mathbf{w}_A = -\mathbf{i}\frac{b}{2} - \mathbf{j}\frac{\sqrt{3}}{2}b \qquad (2\text{-}6a)$$

By similar computations we may show that

$$\mathbf{w}_B = \mathbf{i}\frac{b}{2} - \mathbf{j}\frac{\sqrt{3}}{2}b \qquad (2\text{-}6b)$$

$$\mathbf{w}_C = \mathbf{i}b \qquad (2\text{-}6c)$$

$$\mathbf{w}_D = -\mathbf{w}_A = \mathbf{i}\frac{b}{2} + \mathbf{j}\frac{\sqrt{3}}{2}b \qquad (2\text{-}6d)$$

$$\mathbf{w}_E = -\mathbf{w}_B = -\mathbf{i}\frac{b}{2} + \mathbf{j}\frac{\sqrt{3}}{2}b \qquad (2\text{-}6e)$$

$$\mathbf{w}_F = -\mathbf{w}_C = -\mathbf{i}b \qquad (2\text{-}6f)$$

As an illustration of the utility of this analysis, we calculate the distance from hydrogen atom A' to hydrogen atom C'. The vector drawn from A' to C' (A′C′) is simply the sum of the bond vectors along the route $A'ABCC'$:

$$(\mathbf{A'C'}) = -\mathbf{w}_A + \mathbf{u}_1 + \mathbf{u}_2 + \mathbf{w}_C$$

$$= \mathbf{i}\frac{3}{2}(a + b) + \mathbf{j}\frac{\sqrt{3}}{2}(a + b)$$

The distance from A' to C' is found by taking the square root of the scalar product of (A′C′) with itself to be $\sqrt{3}\,(a+b)$. Other geometrical relationships can be readily calculated by similar procedures.

2-3 Cyclohexane–boat form

A more complicated but more typical example for the determination of molecular geometry is cyclohexane. This molecule exists in two stable conformations, the boat form and the chair form. We consider in this section the boat conformation and in the following section the chair conformation. For the moment we neglect the hydrogen atoms and consider only the geometric

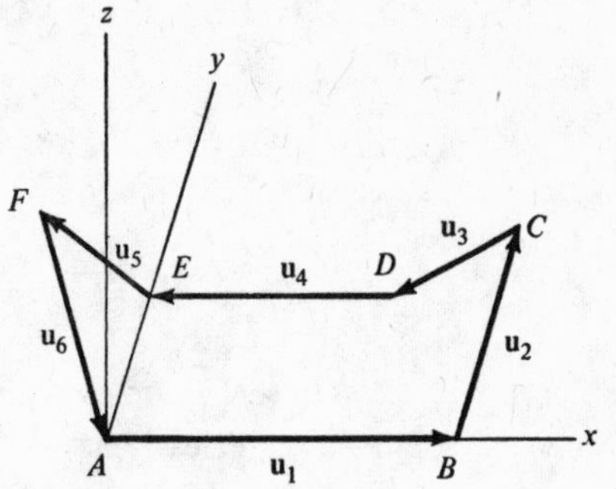

FIGURE 2-4

arrangement of the carbon atoms. The hydrogen atoms for the chair form only are considered in the next section.

The boat conformation of cyclohexane is shown in Fig. 2-4. Atoms *A, B, D,* and *E* lie in a plane. The bonds are all of length *a* and the internal bond angles are all tetrahedral or 109°28′ [$\cos^{-1}(-1/3)$]. An arbitrary choice for the *x, y, z* axes and the bond vectors $\mathbf{u}_1, \mathbf{u}_2, \ldots, \mathbf{u}_6$ are also shown in Fig. 2-4.

The bond vectors $\mathbf{u}_1$ and $\mathbf{u}_4$ are clearly seen to be

$$\mathbf{u}_1 = \mathbf{i}\,a$$

$$\mathbf{u}_4 = -\mathbf{i}\,a$$

To determine the remaining four bond vectors, we first write them in a general form:

$$\mathbf{u}_2 = \mathbf{i}\alpha_x + \mathbf{j}\alpha_y + \mathbf{k}\alpha_z$$
$$\mathbf{u}_3 = \mathbf{i}\beta_x + \mathbf{j}\beta_y + \mathbf{k}\beta_z$$
$$\mathbf{u}_5 = \mathbf{i}\gamma_x + \mathbf{j}\gamma_y + \mathbf{k}\gamma_z$$
$$\mathbf{u}_6 = \mathbf{i}\delta_x + \mathbf{j}\delta_y + \mathbf{k}\delta_z$$

The bond vectors $\mathbf{u}_2, \mathbf{u}_3, \mathbf{u}_5, \mathbf{u}_6$ satisfy the following relationships:

$$-\mathbf{u}_1 \cdot \mathbf{u}_2 = \frac{-a^2}{3} \qquad (2\text{-}7)$$

$$-\mathbf{u}_2 \cdot \mathbf{u}_3 = \frac{-a^2}{3} \qquad (2\text{-}8)$$

$$-\mathbf{u}_3 \cdot \mathbf{u}_4 = \frac{-a^2}{3} \qquad (2\text{-}9)$$

$$-\mathbf{u}_4 \cdot \mathbf{u}_5 = \frac{-a^2}{3} \qquad (2\text{-}10)$$

$$-\mathbf{u}_5 \cdot \mathbf{u}_6 = \frac{-a^2}{3} \qquad (2\text{-}11)$$

$$-\mathbf{u}_6 \cdot \mathbf{u}_1 = \frac{-a^2}{3} \tag{2-12}$$

$$\sum_{i=1}^{6} \mathbf{u}_i = 0 \tag{2-13}$$

$$|\mathbf{u}_i| = a \qquad i = 2, 3, 5, 6 \tag{2-14}$$

Since Eqs. (2-7) to (2-14) are eleven relations for the twelve unknowns $\alpha_x, \ldots, \delta_z$, we shall clearly have to use some knowledge regarding the symmetry of the molecule in order to determine the bond vectors.

Since *A, B, D, E* all lie on the *xy* plane, the *z* component of $\mathbf{u}_2$ is the negative of the *z* component of $\mathbf{u}_3$, and the *z* component of $\mathbf{u}_5$ the negative of that of $\mathbf{u}_6$, so that

$$\alpha_z = -\beta_z \qquad \gamma_z = -\delta_z \tag{2-15}$$

Since atoms *C* and *F* have the same *z* coordinate, we have

$$\alpha_z = \gamma_z \qquad \beta_z = \delta_z \tag{2-16}$$

Combining Eqs. (2-15) and (2-16), we obtain

$$\alpha_z = -\beta_z = \gamma_z = -\delta_z \tag{2-17}$$

Since the *y* coordinates of *A* and *B* are the same ($y = 0$) and the *y* coordinates of *D* and *E* are the same, we have

$$\alpha_y + \beta_y = -(\gamma_y + \delta_y) \tag{2-18}$$

Moreover, since the *y* coordinates of *C* and of *F* are halfway between those of *A* and *E*, we have

$$\alpha_y = \beta_y \qquad \gamma_y = \delta_y \tag{2-19}$$

Equations (2-18) and (2-19) together yield

$$\alpha_y = \beta_y = -\gamma_y = -\delta_y \tag{2-20}$$

Although we could obtain a corresponding relationship among $\alpha_x, \beta_x, \gamma_x, \delta_x$ from symmetry considerations of the *x* components of the bond vectors, we choose instead to utilize Eqs. (2-7) to (2-14). Equation (2-7) gives

$$\mathbf{i}a \cdot (\mathbf{i}\alpha_x + \mathbf{j}\alpha_y + \mathbf{k}\alpha_z) = \frac{a^2}{3}$$

from which we obtain $\alpha_x = a/3$. Similarly, Eqs. (2-9), (2-10), and (2-12) give, respectively, $\beta_x = -a/3$, $\gamma_x = -a/3$, $\delta_x = a/3$.

The remaining problem is that of finding values for α_y and α_z. From Eq. (2-8) we obtain

$$(\mathbf{i}\alpha_x + \mathbf{j}\alpha_y + \mathbf{k}\alpha_z) \cdot (\mathbf{i}\beta_x + \mathbf{j}\beta_y + \mathbf{k}\beta_z) = (\tfrac{a}{3})(-\tfrac{a}{3}) + \alpha_y\beta_y + \alpha_z\beta_z = \frac{a^2}{3}$$

Since $\beta_y = \alpha_y$ and $\beta_z = -\alpha_z$, this expression reduces to

$$\alpha_y{}^2 - \alpha_z{}^2 = \frac{4a^2}{9} \tag{2-21}$$

The same result is obtained from Eq. (2-11). Equation (2-14) for $\mathbf{u}_2$ gives

$$\alpha_x{}^2 + \alpha_y{}^2 + \alpha_z{}^2 = a^2$$

or

$$\alpha_y{}^2 + \alpha_z{}^2 = \frac{8a^2}{9} \tag{2-22}$$

Equations (2-21) and (2-22) are solved simultaneously to yield $\alpha_y = \sqrt{6}\, a/3$ and $\alpha_z = \sqrt{2}\, a/3$, where the positive square roots were taken in agreement with $\mathbf{u}_2$ as shown in Fig. 2-4.

Combining Eqs. (2-17) and (2-20) with the values for $\alpha_x, \beta_x, \gamma_x, \delta_x, \alpha_y, \alpha_z$ obtained above, we can now write explicitly the bond vectors for the boat conformation of cyclohexane:

$$\begin{aligned}
\mathbf{u}_1 &= \mathbf{i}\,a \\
\mathbf{u}_2 &= \mathbf{i}\frac{a}{3} + \mathbf{j}\frac{\sqrt{6}}{3}a + \mathbf{k}\frac{\sqrt{2}}{3}a \\
\mathbf{u}_3 &= -\mathbf{i}\frac{a}{3} + \mathbf{j}\frac{\sqrt{6}}{3}a - \mathbf{k}\frac{\sqrt{2}}{3}a \\
\mathbf{u}_4 &= -\mathbf{i}\,a \\
\mathbf{u}_5 &= -\mathbf{i}\frac{a}{3} - \mathbf{j}\frac{\sqrt{6}}{3}a + \mathbf{k}\frac{\sqrt{2}}{3}a \\
\mathbf{u}_6 &= \mathbf{i}\frac{a}{3} - \mathbf{j}\frac{\sqrt{6}}{3}a - \mathbf{k}\frac{\sqrt{2}}{3}a
\end{aligned} \tag{2-23}$$

At this point it is advisable to check Eqs. (2-23) for accuracy. First, since the bond vectors $\mathbf{u}_1, \ldots, \mathbf{u}_6$ form a cycle, the sum $\mathbf{u}_1 + \mathbf{u}_2 + \mathbf{u}_3 + \mathbf{u}_4 + \mathbf{u}_5 + \mathbf{u}_6$ must be zero. Second, each bond vector $\mathbf{u}_i$ must have a magnitude a. Third, we can readily see in Fig. 2-4 that $\mathbf{u}_1$, $\mathbf{u}_2$, and $\mathbf{u}_6$ have positive x components, and the others have negative x components. Moreover, $\mathbf{u}_2$ and $\mathbf{u}_3$ have positive y components, $\mathbf{u}_5$ and $\mathbf{u}_6$ have negative y components, and $\mathbf{u}_1$ and $\mathbf{u}_4$ have no y components. Likewise, $\mathbf{u}_2$ and $\mathbf{u}_5$ have positive z components, $\mathbf{u}_3$ and $\mathbf{u}_6$ have negative z components, and $\mathbf{u}_1$ and $\mathbf{u}_4$ have no z components. These quick checks are not sufficient to ensure the accuracy of the bond vectors, but they often reveal minor errors.

The bond vectors in Eqs. (2-23) may be used to calculate various geometric properties of the boat form of cyclohexane. For example, the distance (AE) is the magnitude of the vector **(AE)**, so that

$$(\mathbf{AE}) = -\mathbf{u}_6 - \mathbf{u}_5 = \mathbf{j}\frac{2\sqrt{6}}{3}a$$

$$(AE) = |(\mathbf{AE})| = \frac{2\sqrt{6}}{3}a = 1.633a$$

Similarly, the distance (CF) is the magnitude of the vector **(CF)**:

$$(\mathbf{CF}) = \mathbf{u}_3 + \mathbf{u}_4 + \mathbf{u}_5 = \mathbf{i}\frac{5}{3}a$$

$$(CF) = |(\mathbf{CF})| = \frac{5}{3}a = 1.667a$$

A considerably more difficult problem is to find the angle between the plane *ABDE* and the plane *AEF*. To solve this problem we construct the vector **(GF)**, where G is the midpoint of the line (AE), as shown in Fig. 2-5. Note that the line (GF) is perpendicular to (AE). Since the location of the origin of a vector is immaterial, the vector $\mathbf{u}_1$ can be moved so that its origin is at point G (see Fig. 2-5). The new vector $\mathbf{u}_1$ is, of course, parallel to **(AB)** and still has a magnitude a. The angle θ which we wish to find is that between **(GF)** and $\mathbf{u}_1$; thus

$$\cos\theta = \frac{(\mathbf{GF})\cdot\mathbf{u}_1}{|(\mathbf{GF})|\,a}$$

The problem reduces, therefore, to constructing the vector **(GF)**. Since *AGFA* forms a cycle, we have

$$(\mathbf{AG}) + (\mathbf{GF}) + (\mathbf{FA}) = 0$$

The vector **(FA)** is just $\mathbf{u}_6$ in Eqs. (2-23). The vector **(AG)** is one-half **(AE)** or

$$(\mathbf{AG}) = \frac{1}{2}(\mathbf{AE}) = \mathbf{j}\frac{\sqrt{6}}{3}a$$

Thus **(GF)** is found to be

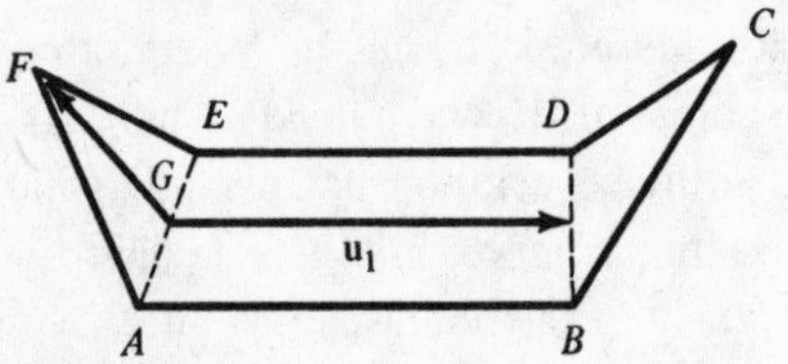

FIGURE 2-5

$$(GF) = -\mathbf{i}\,\frac{1}{3}a + \mathbf{k}\frac{\sqrt{2}}{3}a$$

The magnitude of (GF) is then $\sqrt{3}\,a/3$ and

$$\cos\theta = \frac{-a^2/3}{\sqrt{3}\,a^2/3} = -\frac{\sqrt{3}}{3} = -0.57735$$

$$\theta = \cos^{-1}(-0.57735) = 125°16'$$

2-4 Cyclohexane–chair form

We now consider the chair conformation of cyclohexane as shown in Fig. 2-6. The carbon-carbon bond lengths and the bond angles are the same as those for the boat form. We make the same choice for the cartesian coordinate system and define the bond vectors $\mathbf{v}_1, \mathbf{v}_2, \ldots, \mathbf{v}_6$ in a manner similar to the boat form (see Fig. 2-6). A comparison of Figs. 2-4 and 2-6 reveals that each bond vector $\mathbf{v}_i$ in the chair conformation is equal to its corresponding bond vector $\mathbf{u}_i$ in the boat conformation except that the z components of $\mathbf{v}_2$ and $\mathbf{v}_3$ are the negatives, respectively, of the z components of $\mathbf{u}_2$ and $\mathbf{u}_3$. Thus, we have

$$\begin{aligned}
\mathbf{v}_1 &= \mathbf{i}\,a \\
\mathbf{v}_2 &= \mathbf{i}\frac{a}{3} + \mathbf{j}\frac{\sqrt{6}}{3}a - \mathbf{k}\frac{\sqrt{2}}{3}a \\
\mathbf{v}_3 &= -\mathbf{i}\frac{a}{3} + \mathbf{j}\frac{\sqrt{6}}{3}a + \mathbf{k}\frac{\sqrt{2}}{3}a \\
\mathbf{v}_4 &= -\mathbf{i}\,a \\
\mathbf{v}_5 &= -\mathbf{i}\frac{a}{3} - \mathbf{j}\frac{\sqrt{6}}{3}a + \mathbf{k}\frac{\sqrt{2}}{3}a \\
\mathbf{v}_6 &= \mathbf{i}\frac{a}{3} - \mathbf{j}\frac{\sqrt{6}}{3}a - \mathbf{k}\frac{\sqrt{2}}{3}a
\end{aligned} \qquad (2\text{-}24)$$

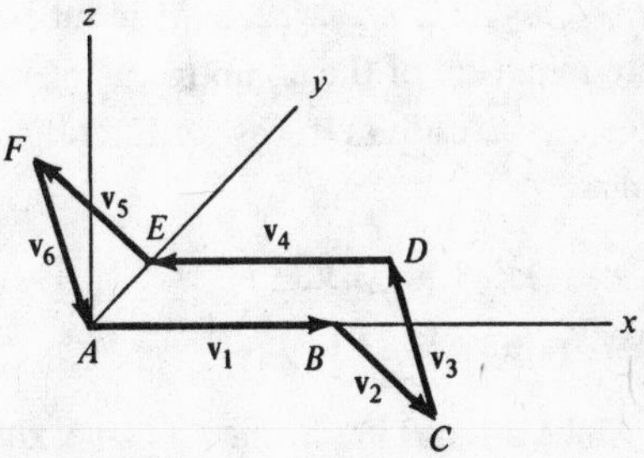

FIGURE 2-6

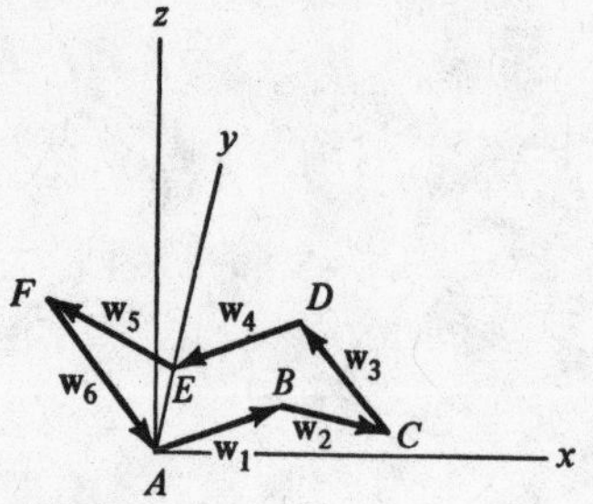

FIGURE 2-7

As an example we calculate the distance (CF) for the chair form. The vector $(\mathbf{CF})$ is given by

$$(\mathbf{CF}) = \mathbf{v}_6 + \mathbf{v}_1 + \mathbf{v}_2 = \mathbf{i}\frac{5}{3}a - \mathbf{k}\frac{2\sqrt{2}}{3}a$$

and its magnitude is

$$(CF) = |(\mathbf{CF})| = \left(\frac{25}{9}a^2 + \frac{8}{9}a^2\right)^{1/2} = \frac{\sqrt{33}}{3}a$$

It is instructive to reexamine at this point the chair conformation of cyclohexane with a different orientation of cartesian coordinate axes. As the xy plane we now choose the plane determined by the three points A, C, and E. The points B, D, and F also determine a plane, which is parallel to the plane ACE. Figure 2-7 depicts this new orientation of the molecule. We again select atom A as the origin and let the y axis pass through atom E. The x axis is, of course, determined by the requirement that atom C lie in the xy plane. The orientation shown in Fig. 2-6 may be transformed to that in Fig. 2-7 by rotating the molecule in a counterclockwise direction about the line (AE) until the point C lies in the xy plane.

To avoid confusion we denote the bond vectors in this new coordinate system by $\mathbf{w}_1, \mathbf{w}_2, \ldots, \mathbf{w}_6$, where the numbering is the same as before. The determination of the components of $\mathbf{w}_1, \ldots, \mathbf{w}_6$ is more complicated for this choice of axes than it was for the original choice. We begin by considering $\mathbf{w}_1$ and $\mathbf{w}_2$:

$$\mathbf{w}_1 = \mathbf{i}\alpha_x + \mathbf{j}\alpha_y + \mathbf{k}\alpha_z$$
$$\mathbf{w}_2 = \mathbf{i}\beta_x + \mathbf{j}\beta_y + \mathbf{k}\beta_z$$

Since atom B lies in the xz plane, the bond vector $\mathbf{w}_1$ also lies in that plane and we have $\alpha_y = 0$. Since atoms A and C lie on the $z = 0$ plane, we must have

$\alpha_z = -\beta_z$. Furthermore, the magnitude of $\mathbf{w}_1$ and $\mathbf{w}_2$ both have a value a:

$$\mathbf{w}_1 \cdot \mathbf{w}_1 = \alpha_x^2 + \alpha_y^2 + \alpha_z^2 = a^2$$
$$\mathbf{w}_2 \cdot \mathbf{w}_2 = \beta_x^2 + \beta_y^2 + \beta_z^2 = a^2$$

Thus we obtain the relations

$$\alpha_z^2 = a^2 - \alpha_x^2$$
$$\begin{aligned} \beta_y^2 &= a^2 - \beta_z^2 - \beta_x^2 \\ &= a^2 - \alpha_z^2 - \beta_x^2 \\ &= \alpha_x^2 - \beta_x^2 \end{aligned}$$

so that at this point $\mathbf{w}_1$ and $\mathbf{w}_2$ have only two unknowns, α_x and β_x:

$$\mathbf{w}_1 = \mathbf{i}\alpha_x + \mathbf{k}(a^2 - \alpha_x^2)^{1/2}$$
$$\mathbf{w}_2 = \mathbf{i}\beta_x + \mathbf{j}(\alpha_x^2 - \beta_x^2)^{1/2} - \mathbf{k}(a^2 - \alpha_x^2)^{1/2}$$

Since $-\mathbf{w}_1$ and $\mathbf{w}_2$ form a tetrahedral angle, we have

$$-\mathbf{w}_1 \cdot \mathbf{w}_2 = -\frac{1}{3}a^2$$

from which it follows that

$$\beta_x = \frac{4a^2}{3\alpha_x} - \alpha_x$$

or

$$\mathbf{w}_2 = \mathbf{i}\left(\frac{4a^2}{3\alpha_x} - \alpha_x\right) + \mathbf{j}\left(\frac{8}{3}a^2 - \frac{16a^4}{9\alpha_x^2}\right)^{1/2} - \mathbf{k}(a^2 - \alpha_x^2)^{1/2}$$

We now consider the bond vector $\mathbf{w}_3$,

$$\mathbf{w}_3 = \mathbf{i}\gamma_x + \mathbf{j}\gamma_y + \mathbf{k}\gamma_z$$

and note that since *A, B, D,* and *E* form a rectangle, we have $\gamma_x = -\beta_x$ and $\gamma_z = -\beta_z$. Moreover, the y coordinate of *C* is half that of *D*, so that we have $\gamma_y = \beta_y$. These arguments lead to

$$\mathbf{w}_3 = \mathbf{i}\left(\frac{4a^2}{3\alpha_x} - \alpha_x\right) + \mathbf{j}\left(\frac{8}{3}a^2 - \frac{16a^4}{9\alpha_x^2}\right)^{1/2} + \mathbf{k}(a^2 - \alpha_x^2)^{1/2}$$

The bond vectors $-\mathbf{w}_2$ and $\mathbf{w}_3$ form a tetrahedral angle,

$$-\mathbf{w}_2 \cdot \mathbf{w}_3 = -\frac{1}{3}a^2$$

which leads to

$$\alpha_x = \frac{2\sqrt{2}}{3}a$$

We further note that $\mathbf{w}_4, \mathbf{w}_5, \mathbf{w}_6$ are antiparallel, respectively, to $\mathbf{w}_1, \mathbf{w}_2$, and $\mathbf{w}_3$. Thus, the final bond vectors are

$$\begin{aligned}
\mathbf{w}_1 &= \mathbf{i}\frac{2\sqrt{2}}{3}a + \mathbf{k}\frac{a}{3} \\
\mathbf{w}_2 &= \mathbf{i}\frac{\sqrt{2}}{3}a + \mathbf{j}\frac{\sqrt{6}}{3}a - \mathbf{k}\frac{a}{3} \\
\mathbf{w}_3 &= -\mathbf{i}\frac{\sqrt{2}}{3}a + \mathbf{j}\frac{\sqrt{6}}{3}a + \mathbf{k}\frac{a}{3} \\
\mathbf{w}_4 &= -\mathbf{i}\frac{2\sqrt{2}}{3}a - \mathbf{k}\frac{a}{3} \\
\mathbf{w}_5 &= -\mathbf{i}\frac{\sqrt{2}}{3}a - \mathbf{j}\frac{\sqrt{6}}{3}a + \mathbf{k}\frac{a}{3} \\
\mathbf{w}_6 &= \mathbf{i}\frac{\sqrt{2}}{3}a - \mathbf{j}\frac{\sqrt{6}}{3}a - \mathbf{k}\frac{a}{3}
\end{aligned} \tag{2-25}$$

This representation of the bond vectors in the chair conformation of cyclohexane is completely equivalent to the representation in Eqs. (2-24). The two sets differ only in a rotation of the arbitrarily chosen set of cartesian coordinate axes.

We now consider the hydrogen atoms in the chair conformation. Each carbon atom has two hydrogen atoms bonded to it with a bond length b such that each carbon-carbon-hydrogen and hydrogen-carbon-hydrogen bond angle is tetrahedral (see Fig. 2-8). We define the carbon-hydrogen bond vectors as directed from the carbon atom to the corresponding hydrogen atom and proceed to evaluate their components in terms of the coordinate system shown in Fig. 2-7.

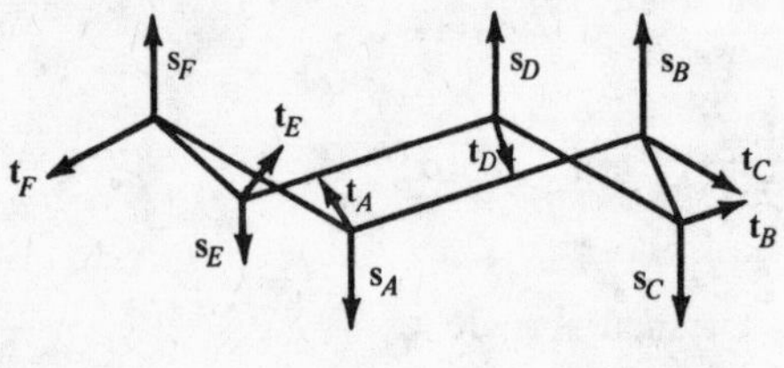

FIGURE 2-8

We consider first the carbon-hydrogen bond vectors on carbon atom A. Let **q** with components q_x, q_y, q_z represent either of them. Since **q** forms tetrahedral angles with $\mathbf{w}_1$ and $-\mathbf{w}_6$, we have

$$\mathbf{w}_1 \cdot \mathbf{q} = -\frac{1}{3}ab$$

$$-\mathbf{w}_6 \cdot \mathbf{q} = -\frac{1}{3}ab$$

from which it follows that

$$q_y = \sqrt{3}\, q_x$$

$$q_z = -2\sqrt{2}\, q_x - b$$

From the relation

$$\mathbf{q} \cdot \mathbf{q} = q_x^2 + q_y^2 + q_z^2 = b^2$$

we obtain two possible solutions:

$$q_x = 0 \qquad q_y = 0 \qquad q_z = -b$$

and

$$q_x = -\frac{\sqrt{2}}{3}b \qquad q_y = -\frac{\sqrt{6}}{3}b \qquad q_z = \frac{b}{3}$$

Let the first set of components be a carbon-hydrogen bond vector $\mathbf{s}_A$ and the second set $\mathbf{t}_A$:

$$\mathbf{s}_A = -\mathbf{k}b$$

$$\mathbf{t}_A = -\mathbf{i}\frac{\sqrt{2}}{3}b - \mathbf{j}\frac{\sqrt{6}}{3}b + \mathbf{k}\frac{b}{3} \qquad (2\text{-}26)$$

We note $\mathbf{s}_A$ and $\mathbf{t}_A$ form a tetrahedral angle as required, i.e.,

$$\mathbf{s}_A \cdot \mathbf{t}_A = -\frac{1}{3}b^2$$

A similar treatment of the carbon-hydrogen bond vectors on the other carbon atoms yields

$$\mathbf{s}_B = \mathbf{k}b \qquad \mathbf{t}_B = \mathbf{i}\frac{\sqrt{2}}{3}b - \mathbf{j}\frac{\sqrt{6}}{3}b - \mathbf{k}\frac{b}{3}$$

$$\mathbf{s}_C = -\mathbf{k}b \qquad \mathbf{t}_C = \mathbf{i}\frac{2\sqrt{2}}{3}b + \mathbf{k}\frac{b}{3}$$

$$\mathbf{s}_D = \mathbf{k}b \qquad \mathbf{t}_D = \mathbf{i}\frac{\sqrt{2}}{3}b + \mathbf{j}\frac{\sqrt{6}}{3}b - \mathbf{k}\frac{b}{3}$$

$$\mathbf{s}_E = -\mathbf{k}b \qquad \mathbf{t}_E = -\mathbf{i}\frac{\sqrt{2}}{3}b + \mathbf{j}\frac{\sqrt{6}}{3}b + \mathbf{k}\frac{b}{3}$$
$$\mathbf{s}_F = \mathbf{k}b \qquad \mathbf{t}_F = -\mathbf{i}\frac{2\sqrt{2}}{3}b - \mathbf{k}\frac{b}{3} \tag{2-27}$$

The carbon-hydrogen bond vectors $\mathbf{s}_A, \ldots, \mathbf{s}_F$ are alternately parallel and antiparallel to the positive z axis for this choice of the cartesian coordinate system. The corresponding bonds are called *axial* bonds. The vectors $\mathbf{t}_A, \ldots, \mathbf{t}_F$ make an angle of 19°28′ with their projections on the xy plane. The corresponding bonds are therefore known as *equatorial* bonds.

2-5 1,2-Dimethylcyclopropane

We consider one more example of the determination of molecular geometry, the cis and trans enantiomers of 1,2-dimethylcyclopropane. The three carbon atoms of the ring are labeled A, B, and C, the methyl groups D and E, and two of the hydrogen atoms F and G, as shown in Fig. 2-9. The two hydrogen atoms on carbon atom C will be ignored here. An arbitrary choice of coordinate axes with A as origin is shown in Fig. 2-9 for the cis enantiomer. The same choice is made for the trans enantiomer. The bond vectors $\mathbf{u}_i$ for the cis and $\mathbf{v}_i$ for the trans enantiomers are defined as shown in Fig. 2-9.

The carbon-carbon bond length in cyclopropane is 1.51 Å.[1] We assume here

[1] O. Bastiansen, F. N. Fritsch, and K. Hedberg, *Acta Cryst.*, **17**: 538 (1963).

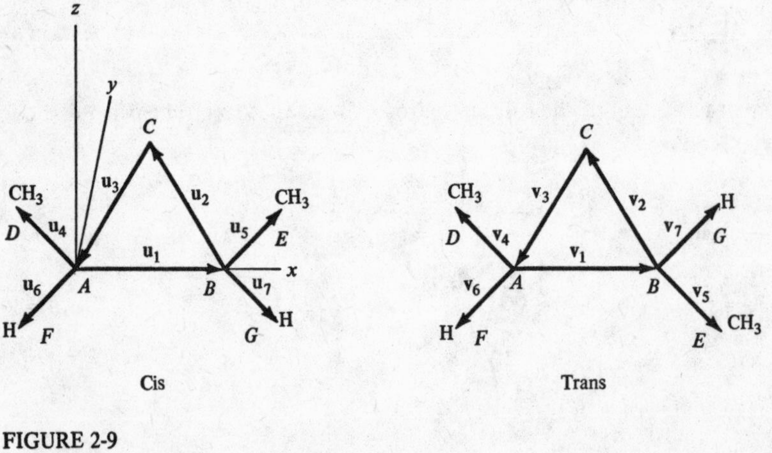

FIGURE 2-9

that the carbon-carbon bond lengths in the ring of 1,2-dimethylcyclopropane are the same as for cyclopropane. Thus we assume the triangle *ABC* to be equilateral and the internal bond angles to be 60°. We further assume that the angles *DAF* and *EBG* are 120° and that these angles are bisected by the plane of the ring. The center of the ring, a carbon atom of the ring, and the centers of the hydrogen atom and the methyl group attached to that carbon atom are assumed to lie in a plane. The bond distance from a ring carbon atom to the carbon atom of its substituent methyl group is taken as 1.540 Å and the carbon-hydrogen bond distances as 1.089 Å [1]

We first treat the cis enantiomer. The bond vectors $\mathbf{u}_1$, $\mathbf{u}_2$, and $\mathbf{u}_3$ associated with the ring are easily seen to be

$$\mathbf{u}_1 = \mathbf{i}\,1.510$$

$$\mathbf{u}_2 = -\mathbf{i}\,1.510 \cos 60^\circ + \mathbf{j}\,1.510 \sin 60^\circ$$
$$= -\mathbf{i}\,0.755 + \mathbf{j}\,0.755\sqrt{3}$$

$$\mathbf{u}_3 = -\mathbf{u}_1 - \mathbf{u}_2 = -\mathbf{i}\,0.755 - \mathbf{j}\,0.755\sqrt{3}$$

The projection of $\mathbf{u}_4$ on the z axis is 1.540 sin 60° and on the xy plane 1.540 cos 60°. The latter projection makes an angle of 30° with the *negative* x axis and 60° with the *negative* y axis. Therefore, we have

$$\mathbf{u}_4 = -\mathbf{i}(1.540 \cos 60^\circ) \cos 30^\circ - \mathbf{j}(1.540 \cos 60^\circ) \cos 60^\circ + \mathbf{k}\,1.540 \sin 60^\circ$$
$$= -\mathbf{i}\,0.385\sqrt{3} - \mathbf{j}\,0.385 + \mathbf{k}\,0.770\sqrt{3}$$

The bond vector $\mathbf{u}_5$ differs from $\mathbf{u}_4$ only in the sign of its x component, so that

$$\mathbf{u}_5 = \mathbf{i}\,0.385\sqrt{3} - \mathbf{j}\,0.385 + \mathbf{k}\,0.770\sqrt{3}$$

The vectors $\mathbf{u}_6$ and $\mathbf{u}_7$ differ from $\mathbf{u}_4$ and $\mathbf{u}_5$, respectively, in the sign of the z components and in magnitude. Since the C–H bond length is 1.089 Å compared with 1.540 Å for the C–CH_3 bond length, $\mathbf{u}_6$ and $\mathbf{u}_7$ are

$$\mathbf{u}_6 = -\mathbf{i}\,0.272\sqrt{3} - \mathbf{j}\,0.272 - \mathbf{k}\,0.545\sqrt{3}$$
$$\mathbf{u}_7 = \mathbf{i}\,0.272\sqrt{3} - \mathbf{j}\,0.272 - \mathbf{k}\,0.545\sqrt{3}$$

In the trans enantiomer, we have

$$\mathbf{v}_1 = \mathbf{u}_1 \qquad \mathbf{v}_4 = \mathbf{u}_4$$
$$\mathbf{v}_2 = \mathbf{u}_2 \qquad \mathbf{v}_6 = \mathbf{u}_6$$
$$\mathbf{v}_3 = \mathbf{u}_3$$

[1] O. Bastiansen, F. N. Fritsch, and K. Hedberg, *Acta Cryst.*, **17**:538 (1963).

The vectors $\mathbf{u}_5$ and $\mathbf{v}_5$ are identical except for the sign of their z components, so that

$$\mathbf{v}_5 = \mathbf{i}0.385\sqrt{3} - \mathbf{j}0.385 - \mathbf{k}0.770\sqrt{3}$$

Similarly, $\mathbf{u}_7$ and $\mathbf{v}_7$ differ only in the sign of their z components,

$$\mathbf{v}_7 = \mathbf{i}0.272\sqrt{3} - \mathbf{j}0.272 + \mathbf{k}0.545\sqrt{3}$$

As an example, we now determine the angle between the two methyl groups in each of the two enantiomers. In terms of the bond vectors, we want the angles between $\mathbf{u}_4$ and $\mathbf{u}_5$ and between $\mathbf{v}_4$ and $\mathbf{v}_5$. For the cis enantiomer, we have

$$\mathbf{u}_4 \cdot \mathbf{u}_5 = (1.540)^2 \cos\theta$$

The dot product is

$$\begin{aligned}\mathbf{u}_4 \cdot \mathbf{u}_5 &= (-\mathbf{i}0.385\sqrt{3} - \mathbf{j}0.385 + \mathbf{k}0.770\sqrt{3}) \cdot (\mathbf{i}0.385\sqrt{3} - \mathbf{j}0.385 \\ &\quad + \mathbf{k}0.770\sqrt{3}) \\ &= -3(0.385)^2 + (0.385)^2 + 3(0.770)^2 \\ &= (5/8)(1.540)^2\end{aligned}$$

Thus we have $\cos\theta = 5/8$ and $\theta = 51°20'$. For the trans enantiomer,

$$\mathbf{v}_4 \cdot \mathbf{v}_5 = (1.540)^2 \cos\theta$$

and $\cos\theta = -\frac{7}{8}$ or $\theta = 151°$.

2-6 Unit-cell calculations

In crystal lattices the smallest repeating unit is called a *unit cell*. Such a unit cell is described by a parallelepiped, which when repeated throughout the crystal fills all space, leaving no empty regions. The size and shape of a unit cell are specified by the lengths a, b, and c of the three independent edges and the three angles α, β, and γ between each pair of edges (see Fig. 2-10). With one corner of the unit cell as origin, it is convenient in the treatment of crystal structures to select a coordinate system with the edges a, b, and c as axes rather than to use a cartesian coordinate system. Except for a cubic crystal, in which $\alpha = \beta = \gamma = 90°$, the crystal coordinate axes are not orthogonal.

The position of a point P within a unit cell may be specified by three coordinates ξ, η, ζ of the crystal coordinate system. The point ξ, η, ζ is determined by traveling a distance ξa from the origin along the a axis, then a distance ηb parallel to the b axis, and finally a distance ζc parallel to the c axis. If a coordinate, ξ for example, is less than unity, the point P is within the unit cell. If ξ is greater than 1 but less than 2, P is in the adjacent unit cell. For

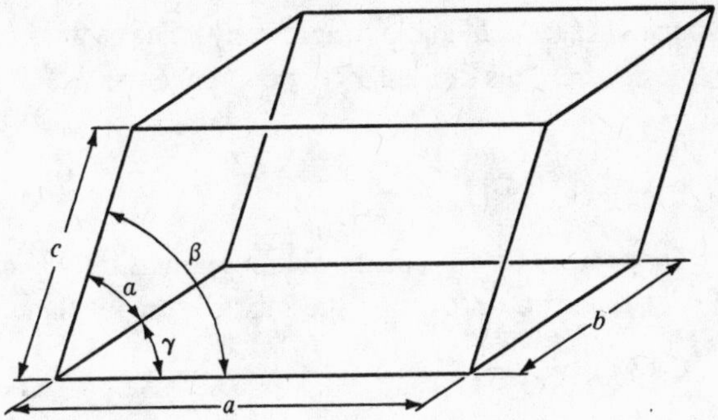

FIGURE 2-10

$\xi > 2$, P lies within a more remote unit cell. Similar conditions hold for η and ζ and the replication of the unit cells in the b and c directions, respectively. Thus, the advantage of the crystal coordinate system over the cartesian coordinate system is apparent: Two points in a crystal are equivalent if the fractional parts of their coordinates ξ, η, ζ are equal. For example, the points (0.80, 0.20, 0.50) and (1.80, 0.20, 0.50) are equivalent but are located in adjacent unit cells.

Calculations involving nonorthogonal coordinate systems are more difficult than those with cartesian coordinates. In fact, some standard, frequently used expressions for unit cells are best obtained by means of cartesian vectors. As an illustration of such an application of vector algebra, we determine in this section expressions for the volume V of a unit cell and the distance l between any two points in a crystal in terms of the unit-cell parameters $a, b, c, \alpha, \beta, \gamma$.

We select a right-handed cartesian coordinate system, as shown in Fig. 2-11, with the x axis along the edge a and the y axis in the plane determined by the edges a and b. Let **A**, **B**, and **C** be vectors directed away from the origin and

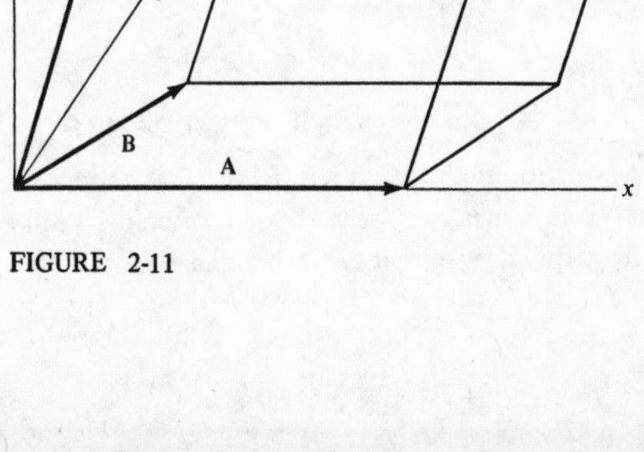

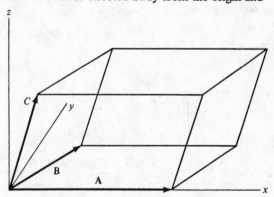

FIGURE 2-11

along the edges a, b, and c, respectively. The vectors **A** and **B** are readily expressed in terms of their x, y, z components:

$$\mathbf{A} = \mathbf{i}a \tag{2-28}$$

$$\mathbf{B} = \mathbf{i}b \cos\gamma + \mathbf{j}b \sin\gamma \tag{2-29}$$

The vector **C** is difficult to obtain geometrically because the angles α and β do not project **C** onto either the z axis or the xy plane. Accordingly, we let

$$\mathbf{C} = \mathbf{i}C_x + \mathbf{j}C_y + \mathbf{k}C_z \tag{2-30}$$

and note that

$$\mathbf{A} \cdot \mathbf{C} = ac \cos\beta \tag{2-31}$$

$$\mathbf{B} \cdot \mathbf{C} = bc \cos\alpha \tag{2-32}$$

$$\mathbf{C} \cdot \mathbf{C} = c^2 \tag{2-33}$$

From Eq. (2-31), we find that

$$C_x = c \cos\beta \tag{2-34}$$

Equations (2-32) and (2-34) yield

$$C_y = \frac{c(\cos\alpha - \cos\beta\cos\gamma)}{\sin\gamma} \tag{2-35}$$

and Eqs. (2-33) to (2-35) give

$$\begin{aligned} C_z &= \frac{c}{\sin\gamma}(\sin^2\gamma - \cos^2\alpha - \cos^2\beta + 2\cos\alpha\cos\beta\cos\gamma)^{1/2} \\ &= \frac{c}{\sin\gamma}(1 - \cos^2\alpha - \cos^2\beta - \cos^2\gamma + 2\cos\alpha\cos\beta\cos\gamma)^{1/2} \end{aligned} \tag{2-36}$$

where the positive square root applies.

The volume V of the unit cell is given by the triple scalar product $\mathbf{A} \cdot \mathbf{B} \times \mathbf{C}$ as expressed in Eq. (1-31):

$$\begin{aligned} V = \mathbf{A} \cdot \mathbf{B} \times \mathbf{C} &= \begin{vmatrix} a & 0 & 0 \\ b\cos\gamma & b\sin\gamma & 0 \\ C_x & C_y & C_z \end{vmatrix} \\ &= ab \sin\gamma\, C_z \\ &= abc\,(1 - \cos^2\alpha - \cos^2\beta - \cos^2\gamma + 2\cos\alpha\cos\beta\cos\gamma)^{1/2} \end{aligned} \tag{2-37}$$

We can also use the vectors **A**, **B**, and **C** as expressed in cartesian coordinates to find the distance l between any pair of points within a crystal. Both points may lie within a given unit cell or they may be in different unit cells. In general, the point P or (ξ, η, ζ) may be located by the vector **P**:

$$\mathbf{P} = \xi\mathbf{A} + \eta\mathbf{B} + \zeta\mathbf{C}$$

Since **A**, **B**, and **C** are constant vectors, the vector **l** from an arbitrary point: P_1 or (ξ_1, η_1, ζ_1) to another arbitrary point P_2 or (ξ_2, η_2, ζ_2) is

$$\mathbf{l} = \mathbf{P}_2 - \mathbf{P}_1 = (\xi_2 - \xi_1)\mathbf{A} + (\eta_2 - \eta_1)\mathbf{B} + (\zeta_2 - \zeta_1)\mathbf{C}$$

The vector **l** may be expressed in terms of cartesian coordinates through Eqs. (2-28) to (2-30):

$$\begin{aligned}\mathbf{l} &= \mathbf{i}\,a(\xi_2 - \xi_1) + (\mathbf{i}\,b \cos\gamma + \mathbf{j} \sin\gamma)(\eta_2 - \eta_1) + (\mathbf{i}\,C_x + \mathbf{j}\,C_y + \mathbf{k}\,C_z)(\zeta_2 - \zeta_1)\\ &= \mathbf{i}[a(\xi_2 - \xi_1) + b\cos\gamma\,(\eta_2 - \eta_1) + C_x(\zeta_2 - \zeta_1)]\\ &\quad + \mathbf{j}[b \sin\gamma(\eta_2 - \eta_1) + C_y(\zeta_2 - \zeta_1)] + \mathbf{k}\,C_z(\zeta_2 - \zeta_1)\end{aligned}$$

The distance between P_1 and P_2 is just the magnitude of **l**. Thus we compute $\mathbf{l} \cdot \mathbf{l}$ and take the square root to obtain

$$\begin{aligned}l = [a^2 (\xi_2 - \xi_1)^2 &+ b^2(\eta_2 - \eta_1)^2 + c^2(\zeta_2 - \zeta_1)^2\\ &+ 2ab \cos\gamma\,(\xi_2 - \xi_1)(\eta_2 - \eta_1) + 2ac \cos\beta\,(\xi_2 - \xi_1)(\zeta_2 - \zeta_1)\\ &+ 2bc \cos\alpha\,(\eta_2 - \eta_1)(\zeta_2 - \zeta_1)]^{1/2}\end{aligned} \tag{2-38}$$

where we have used Eqs. (2-34) to (2-36) and the fact that $\sin^2\gamma + \cos^2\gamma = 1$.

PROBLEMS

1. Determine the bond vectors for benzene with the origin for the cartesian coordinate system at atom F in Fig. 2-2. Calculate the distances (AC), (AD), and (FC) and the angles DAC and ACD with this choice for origin.

2. Show that the components of $\mathbf{w}_B, \ldots, \mathbf{w}_F$ are those given by Eqs. (2-6*b*) to (2-6*f*).

3. Determine the distance between adjacent hydrogen atoms in benzene.

4. Find the distance between atoms A and C in Fig. 2-4 for the boat conformation of cyclohexane.

5. Find the distance between atoms C and F in Fig. 2-7 for the chair conformation of cyclohexane, using the set of bond vectors given by Eqs. (2-25).

6. Find the distance between a pair of axial hydrogen atoms on the same side of the ring for the chair conformation of cyclohexane.

7. Find the distance between the equatorial hydrogen atoms bonded to adjacent carbon atoms for the chair conformation of cyclohexane.

8. Find the distance between the methyl carbon atoms in each enantiomer of 1, 2-dimethylcyclopropane.

9. Determine a set of bond vectors for the carbon-carbon bonds in the cis and trans enantiomers of 2-butene. Find the distance between the terminal carbon atoms for each enantiomer.

Chapter three

Vector Calculus

3-1 Differentiation of a vector

If to each value of the real scalar variable t in an interval $t_1 \leqslant t \leqslant t_2$ there is assigned a vector **A**, then **A**(t) is a *vector function* of the variable t over that interval. In terms of the components of **A**(t), we may write

$$\mathbf{A}(t) = \mathbf{i}A_x(t) + \mathbf{j}A_y(t) + \mathbf{k}A_z(t) \tag{3-1}$$

The derivative of a vector **A**(t) with respect to the variable t is defined by

$$\frac{d\mathbf{A}}{dt} = \lim_{\Delta t \to 0} \frac{\mathbf{A}(t+\Delta t) - \mathbf{A}(t)}{\Delta t} \tag{3-2}$$

Since division of a vector by a scalar yields another vector, the derivative $d\mathbf{A}/dt$ is also a vector quantity. Differentiation of Eq. (3-1) with respect to t gives

$$\frac{d\mathbf{A}}{dt} = \mathbf{i}\frac{dA_x}{dt} + \mathbf{j}\frac{dA_y}{dt} + \mathbf{k}\frac{dA_z}{dt} \tag{3-3}$$

In general, the ratio $(dA_x/dt) : (dA_y/dt) : (dA_z/dt)$ differs from the ratio $A_x : A_y : A_z$, so that the vector $d\mathbf{A}/dt$ is not in the same direction as **A**.

In view of the definition (3-2) and the relationship (3-3), the rules of differential calculus for scalars can be readily extended to vector sums and products. Thus, if $\phi(t)$, $\mathbf{A}(t)$, $\mathbf{B}(t)$ are differentiable functions of t, then

$$\frac{d}{dt}(\mathbf{A} + \mathbf{B}) = \frac{d\mathbf{A}}{dt} + \frac{d\mathbf{B}}{dt} \tag{3-4}$$

$$\frac{d}{dt}(\phi \mathbf{A}) = \phi \frac{d\mathbf{A}}{dt} + \mathbf{A}\frac{d\phi}{dt} \tag{3-5}$$

$$\frac{d}{dt}(\mathbf{A} \cdot \mathbf{B}) = \mathbf{A} \cdot \frac{d\mathbf{B}}{dt} + \mathbf{B} \cdot \frac{d\mathbf{A}}{dt} \tag{3-6}$$

$$\frac{d}{dt}(\mathbf{A} \times \mathbf{B}) = \mathbf{A} \times \frac{d\mathbf{B}}{dt} + \frac{d\mathbf{A}}{dt} \times \mathbf{B} \tag{3-7}$$

The derivative of a vector $\mathbf{A}(t)$ which has a constant magnitude but a changing direction is a vector perpendicular to **A**. To demonstrate this statement, let **A** be a vector of constant magnitude. Then $\mathbf{A} \cdot \mathbf{A} = A^2$ is a constant and

$$\frac{dA^2}{dt} = \frac{d(\mathbf{A} \cdot \mathbf{A})}{dt} = 2\mathbf{A} \cdot \frac{d\mathbf{A}}{dt} = 0 \tag{3-8}$$

If neither **A** nor $d\mathbf{A}/dt$ is zero, then from Eq. (3-8) $d\mathbf{A}/dt$ is perpendicular to **A**.

The second derivative of a vector function $\mathbf{A}(t)$ is defined as the derivative of the derivative:

$$\frac{d^2\mathbf{A}}{dt^2} = \frac{d}{dt}\left(\frac{d\mathbf{A}}{dt}\right) = \mathbf{i}\frac{d^2A_x}{dt^2} + \mathbf{j}\frac{d^2A_y}{dt^2} + \mathbf{k}\frac{d^2A_z}{dt^2} \tag{3-9}$$

Higher derivatives are defined in an analogous fashion.

A vector may be a function of more than one variable. For example, the vector function **A** may depend on its position x, y, z in space, as in the illustration in Sec. 1-1. In this case, we have

$$\mathbf{A}(x,y,z) = \mathbf{i}\,A_x(x,y,z) + \mathbf{j}\,A_y(x,y,z) + \mathbf{k}\,A_z(x,y,z) \tag{3-10}$$

The space derivatives of the components of $A(x,y,z)$ are assumed to exist:

$$\begin{matrix} \dfrac{\partial A_x}{\partial x} & \dfrac{\partial A_y}{\partial x} & \dfrac{\partial A_z}{\partial x} \\[2ex] \dfrac{\partial A_x}{\partial y} & \dfrac{\partial A_y}{\partial y} & \dfrac{\partial A_z}{\partial y} \\[2ex] \dfrac{\partial A_x}{\partial z} & \dfrac{\partial A_y}{\partial z} & \dfrac{\partial A_z}{\partial z} \end{matrix}$$

Here and throughout the remainder of the book, when there is no ambiguity we omit in the partial derivatives the subscripts which indicate the variables being held constant.

A more general case is the situation where the vector function **A** depends on both the space variables x, y, z and the time t. Thus, each component of **A** depends on x, y, z and t, so that

$$\mathbf{A}(x,y,z,t) = \mathbf{i}\,A_x(x,y,z,t) + \mathbf{j}\,A_y(x,y,z,t) + \mathbf{k}\,A_z(x,y,z,t) \tag{3-11}$$

3-2 Vectors in classical mechanics

Of great importance in the development of classical mechanics is the representation of the motion of particles by means of vectors.

One of the basic postulates of classical mechanics is the existence of a fixed set of coordinate axes in space with an origin and unit vectors **i**, **j**, and **k**. The position of a particle with respect to the coordinate axes is denoted by the vector $\mathbf{r} = \mathbf{i}\,x + \mathbf{j}\,y + \mathbf{k}\,z$, which in general is a function of time t:

$$\mathbf{r}(t) = \mathbf{i}\,x(t) + \mathbf{j}\,y(t) + \mathbf{k}\,z(t) \tag{3-12}$$

The velocity vector **v** and the acceleration vector **a** are defined as the first and second derivatives, respectively, of $\mathbf{r}(t)$ with respect to time:

$$\mathbf{v} \equiv \frac{d\mathbf{r}}{dt} = \mathbf{i}\frac{dx}{dt} + \mathbf{j}\frac{dy}{dt} + \mathbf{k}\frac{dz}{dt} \tag{3-13}$$

$$\mathbf{a} \equiv \frac{d\mathbf{v}}{dt} = \frac{d^2\mathbf{r}}{dt^2} = \mathbf{i}\frac{d^2x}{dt^2} + \mathbf{j}\frac{d^2y}{dt^2} + \mathbf{k}\frac{d^2z}{dt^2} \tag{3-14}$$

The total force **F** acting on a particle is the vector sum of all the individual forces acting on the particle and has the form $\mathbf{F} = \mathbf{i}\,F_x + \mathbf{j}\,F_y + \mathbf{k}\,F_z$. A basic equation of classical mechanics is Newton's second law of motion,

$$\mathbf{F} = m\mathbf{a} \tag{3-15}$$

where m is the mass of the particle.

If there is no force acting on a particle, then Eq. (3-15) becomes $m\mathbf{a} = 0$ or

$$\frac{d^2\mathbf{r}}{dt^2} = 0 \tag{3-16}$$

This expression is equivalent to

$$\frac{d^2x}{dt^2} = 0 \qquad \frac{d^2y}{dt^2} = 0 \qquad \frac{d^2z}{dt^2} = 0 \tag{3-17}$$

which yield upon integration

$$x = A + at \qquad y = B + bt \qquad z = C + ct \tag{3-18}$$

where A, B, C, a, b, c are arbitrary constants. In vector form, Eqs. (3-18) are

$$\mathbf{r} = \mathbf{R} + \mathbf{V}t \tag{3-19}$$

where **R** and **V** are arbitrary constant vectors. Thus, the particle moves in a straight line at constant velocity $d\mathbf{r}/dt = \mathbf{V}$. This result is Newton's first law of motion: In the absence of an external force, a particle moves in a straight line with constant velocity.

3-3 Gradient of a scalar field

Let $\phi(x,y,z)$ be a continuous and differentiable scalar function of position in space. The space derivatives of $\phi\,(x,y,z)$ in cartesian coordinates are $\partial\phi/\partial x$, $\partial\phi/\partial y$, $\partial\phi/\partial z$. The *gradient* of $\phi\,(x,y,z)$, often written grad ϕ, is defined as a vector with components $\partial\phi/\partial x$, $\partial\phi/\partial y$, $\partial\phi/\partial z$:

$$\text{grad}\,\phi = \mathbf{i}\frac{\partial\phi}{\partial x} + \mathbf{j}\frac{\partial\phi}{\partial y} + \mathbf{k}\frac{\partial\phi}{\partial z} \tag{3-20}$$

It is convenient to define a vector differential operator ∇ (often called *del*) as

$$\nabla = \mathbf{i}\frac{\partial}{\partial x} + \mathbf{j}\frac{\partial}{\partial y} + \mathbf{k}\frac{\partial}{\partial z} \tag{3-21}$$

When ∇ operates on the scalar function $\phi(x,y,z)$, the result $\nabla\phi$ is just Eq. (3-20), so that

$$\nabla\phi = \text{grad}\,\phi = \mathbf{i}\frac{\partial\phi}{\partial x} + \mathbf{j}\frac{\partial\phi}{\partial y} + \mathbf{k}\frac{\partial\phi}{\partial z} \tag{3-22}$$

We denote by $d\mathbf{r}$ the differential vector displacement with components dx, dy, dz:

$$d\mathbf{r} = \mathbf{i}\,dx + \mathbf{j}\,dy + \mathbf{k}\,dz \tag{3-23}$$

From Eqs. (3-22) and (3-23), we have

$$(\nabla\phi)\cdot d\mathbf{r} = \frac{\partial\phi}{\partial x}dx + \frac{\partial\phi}{\partial y}dy + \frac{\partial\phi}{\partial z}dz \tag{3-24}$$

The right-hand side of Eq. (3-24), however, is just the total differential $d\phi$ of $\phi(x,y,z)$. Thus, the total differential $d\phi$ is the scalar product of $\nabla\phi$ and $d\mathbf{r}$:

$$d\phi = (\nabla\phi)\cdot d\mathbf{r} \tag{3-25}$$

The equation

$$\phi\,(x,y,z) = \text{const} \tag{3-26}$$

represents a surface in cartesian space. As the value of the constant changes, a family of surfaces is obtained. Two members S_1 and S_2 of such a family of surfaces are shown in Fig. 3-1.

Now consider a differential vector displacement $d\mathbf{r}$ which is tangential to the surface S_1 at some point on S_1 (see Fig. 3-1*a*). Since ϕ is constant on S_1, $d\phi$ is

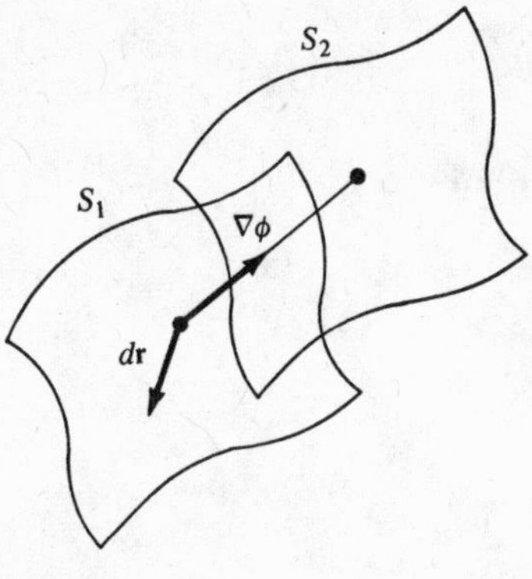

FIGURE 3-1*a*

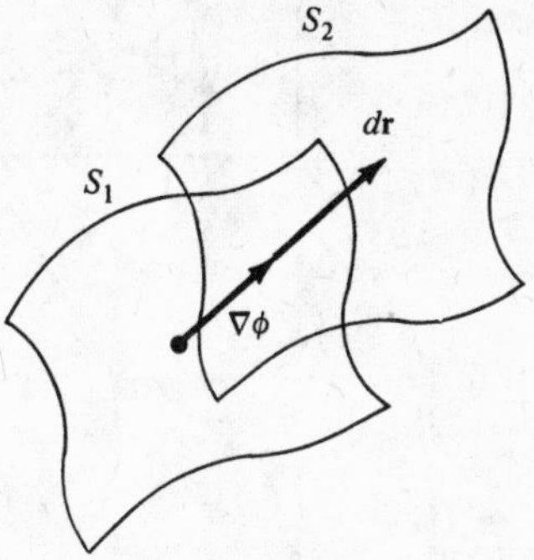

FIGURE 3-1*b*

zero and from Eq. (3-25) $\nabla\phi$ is perpendicular to $d\mathbf{r}$. Thus, $\nabla\phi$ is normal to the surface $\phi = \text{const}$.

The maximum value of $d\phi$ occurs when $\nabla\phi$ and $d\mathbf{r}$ are in the same direction (see Fig. 3-1*b*) for then the scalar product $\nabla\phi \cdot d\mathbf{r}$ in Eq. (3-25) is a maximum. Thus we have

$$(d\phi)_{\max} = |\nabla\phi|\, dr \tag{3-27}$$

or

$$|\nabla\phi| = \left(\frac{d\phi}{dr}\right)_{\max} \tag{3-28}$$

Therefore, $\nabla\phi$ is the direction of the greatest change in ϕ. Moreover, this direction of greatest change is normal to the surface $\phi = \text{const}$ at each point in space.

3-4 Divergence of a vector field

The scalar product of the vector operator ∇ and a vector function $\mathbf{A}(x,y,z)$ is a scalar field which is called the *divergence* of $\mathbf{A}$:

$$\nabla \cdot \mathbf{A} = \frac{\partial A_x}{\partial x} + \frac{\partial A_y}{\partial y} + \frac{\partial A_z}{\partial z} \tag{3-29}$$

The divergence of $\mathbf{A}$ is also often written div $\mathbf{A}$.

If $\phi(x,y,z)$, $\mathbf{A}(x,y,z)$, and $\mathbf{B}(x,y,z)$ are differentiable functions of the space variables, we may obtain from the relation (3-29).

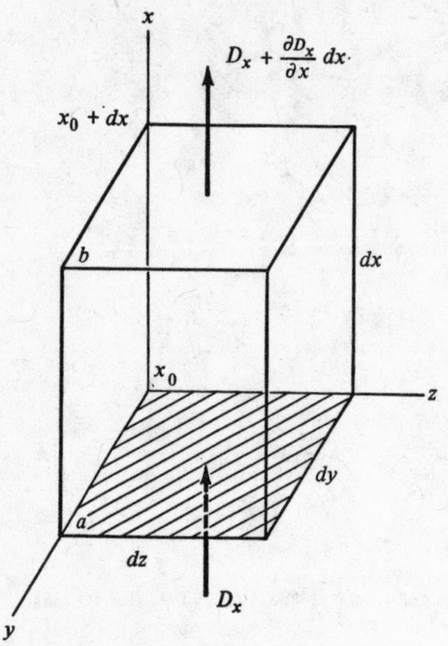

FIGURE 3-2

$$\nabla \cdot (\mathbf{A} + \mathbf{B}) = \nabla \cdot \mathbf{A} + \nabla \cdot \mathbf{B} \tag{3-30}$$

$$\begin{aligned} \nabla \cdot (\phi \mathbf{A}) &= \frac{\partial \phi A_x}{\partial y} + \frac{\partial \phi A_y}{\partial y} + \frac{\partial \phi A_z}{\partial z} \\ &= \phi\left(\frac{\partial A_x}{\partial x} + \frac{\partial A_y}{\partial y} + \frac{\partial A_z}{\partial z}\right) + A_x \frac{\partial \phi}{\partial x} + A_y \frac{\partial \phi}{\partial y} + A_z \frac{\partial \phi}{\partial z} \\ &= \phi \nabla \cdot \mathbf{A} + \mathbf{A} \cdot \nabla \phi \end{aligned} \tag{3-31}$$

The divergence of the gradient of a scalar function is

$$\nabla \cdot \nabla \phi = \frac{\partial^2 \phi}{\partial x^2} + \frac{\partial^2 \phi}{\partial y^2} + \frac{\partial^2 \phi}{\partial z^2} \tag{3-32}$$

The operator $\nabla \cdot \nabla$ is usually abbreviated ∇^2 and is called the laplacian operator:

$$\nabla^2 = \frac{\partial^2}{\partial x^2} + \frac{\partial^2}{\partial y^2} + \frac{\partial^2}{\partial z^2} \tag{3-33}$$

The physical meaning of the divergence of a vector function may be illustrated by considering the flow of a fluid. Consider a small fixed rectangular parallelepiped of dimensions dx, dy, dz as shown in Fig. 3-2. Let the vector function $\mathbf{D}(x, y, z)$ be the mass of fluid flowing through a unit area normal to the direction of flow in unit time. Thus $\mathbf{D}$ is the rate of flow per unit area. The component D_x is the rate of flow per unit area in the x direction through the area $dy\,dz$ at the position x_0 (the shaded area in Fig. 3-2). Similar interpretations hold for the other two components of $\mathbf{D}$.

Let dQ_a be the mass of fluid flowing into the volume element per unit time through the surface at x_0 of area $dy\,dz$ (the shaded area in Fig. 3-2). Therefore, we have

$$dQ_a = D_x\,dy\,dz \tag{3-34}$$

If D_x changes between x_0 and $x_0 + dx$, then the rate of flow per unit area through $dy\,dz$ at the position $x_0 + dx$ may be expanded in a Taylor series:

$$D_x(x_0 + dx) = D_x(x_0) + \left(\frac{\partial D_x}{\partial x}\right)_{x_0} dx + \cdots \tag{3-35}$$

The mass of fluid, dQ_b, leaving the volume element per unit time at $x_0 + dx$ is

$$dQ_b = \left(D_x + \frac{\partial D_x}{\partial x}dx\right)dy\,dz \tag{3-36}$$

where the higher-order terms in the Taylor series expansion [indicated by dots in Eq. (3-35)] have been neglected. The net outward rate of flow through the two faces is simply $dQ_b - dQ_a$ or

$$\left(D_x + \frac{\partial D_x}{\partial x}dx\right)dy\,dz - D_x\,dy\,dz = \frac{\partial D_x}{\partial x}dv \tag{3-37}$$

where $dv = dx\,dy\,dz$ is the volume of the rectangular parallelepiped.

In a similar manner, the net outward flow rate in the y direction can be shown to be

$$\frac{\partial D_y}{\partial y}dv$$

and in the z direction,

$$\frac{\partial D_z}{\partial z}dv$$

The total outward rate of flow dQ for the fluid is the sum of the three components:

$$dQ = \left(\frac{\partial D_x}{\partial x} + \frac{\partial D_y}{\partial y} + \frac{\partial D_z}{\partial z}\right)dv = \nabla \cdot \mathbf{D}\,dv \tag{3-38}$$

Thus, we see that $\nabla \cdot \mathbf{D}$ is the net outward rate of flow per unit volume.

A vector field **A** for which $\nabla \cdot \mathbf{A} = 0$ at all points in space is said to be solenoidal. In the example above, if the fluid is incompressible, it must flow out of the volume element dv at the same rate as it flows in. Thus the outward flow rate vanishes, $\nabla \cdot \mathbf{D}$ is zero, and **D** is solenoidal.

3-5 Curl of a vector field

The cross product of the vector operator ∇ and a vector function $\mathbf{A}(x, y, z)$ is a vector field which is called the *curl* or *rotation* of $\mathbf{A}$:

$$\nabla \times \mathbf{A} = \left(\mathbf{i}\frac{\partial}{\partial x} + \mathbf{j}\frac{\partial}{\partial y} + \mathbf{k}\frac{\partial}{\partial z}\right) \times (\mathbf{i}A_x + \mathbf{j}A_y + \mathbf{k}A_z)$$
$$= \mathbf{i}\left(\frac{\partial A_z}{\partial y} - \frac{\partial A_y}{\partial z}\right) + \mathbf{j}\left(\frac{\partial A_x}{\partial z} - \frac{\partial A_z}{\partial x}\right) + \mathbf{k}\left(\frac{\partial A_y}{\partial x} - \frac{\partial A_x}{\partial y}\right) \tag{3-39}$$

Sometimes the designation curl $\mathbf{A}$ or rot $\mathbf{A}$ is used instead of $\nabla \times \mathbf{A}$. The curl may be written in determinantal form, similar to the vector product (1-27):

$$\nabla \times \mathbf{A} = \begin{vmatrix} \mathbf{i} & \mathbf{j} & \mathbf{k} \\ \frac{\partial}{\partial x} & \frac{\partial}{\partial y} & \frac{\partial}{\partial z} \\ A_x & A_y & A_z \end{vmatrix} \tag{3-40}$$

If $\phi(x,y,z)$, $\mathbf{A}(x,y,z)$, and $\mathbf{B}(x,y,z)$ are differentiable functions of x, y, z, then we have

$$\nabla \times (\mathbf{A} + \mathbf{B}) = (\nabla \times \mathbf{A}) + (\nabla \times \mathbf{B}) \tag{3-41}$$

$$\nabla \times (\phi\mathbf{A}) = \mathbf{i}\left(\frac{\partial \phi A_z}{\partial y} - \frac{\partial \phi A_y}{\partial z}\right) + \mathbf{j}\left(\frac{\partial \phi A_x}{\partial z} - \frac{\partial \phi A_z}{\partial x}\right) + \mathbf{k}\left(\frac{\partial \phi A_y}{\partial x} - \frac{\partial \phi A_x}{\partial y}\right)$$
$$= \phi(\nabla \times \mathbf{A}) + (\nabla\phi) \times \mathbf{A} \tag{3-42}$$

The divergence of the curl of a vector function $\mathbf{A}$ is identically zero:

$$\nabla \cdot \nabla \times \mathbf{A} = \left(\mathbf{i}\frac{\partial}{\partial x} + \mathbf{j}\frac{\partial}{\partial y} + \mathbf{k}\frac{\partial}{\partial z}\right) \cdot \left[\mathbf{i}\left(\frac{\partial A_z}{\partial y} - \frac{\partial A_y}{\partial z}\right) + \mathbf{j}\left(\frac{\partial A_x}{\partial z} - \frac{\partial A_z}{\partial x}\right) + \mathbf{k}\left(\frac{\partial A_y}{\partial x} - \frac{\partial A_x}{\partial y}\right)\right]$$
$$= \left(\frac{\partial^2 A_z}{\partial x\, \partial y} - \frac{\partial^2 A_z}{\partial y\, \partial x}\right) + \left(\frac{\partial^2 A_x}{\partial y\, \partial z} - \frac{\partial^2 A_x}{\partial z\, \partial y}\right) + \left(\frac{\partial^2 A_y}{\partial z\, \partial x} - \frac{\partial^2 A_y}{\partial x\, \partial z}\right) \tag{3-43}$$
$$= 0$$

Moreover, the curl of the gradient of a scalar field vanishes identically:

$$\nabla \times \nabla\phi = \mathbf{i}\left(\frac{\partial^2 \phi}{\partial y\, \partial z} - \frac{\partial^2 \phi}{\partial z\, \partial y}\right) + \mathbf{j}\left(\frac{\partial^2 \phi}{\partial z\, \partial x} - \frac{\partial^2 \phi}{\partial x\, \partial z}\right) + \mathbf{k}\left(\frac{\partial^2 \phi}{\partial x\, \partial y} - \frac{\partial^2 \phi}{\partial y\, \partial x}\right)$$
$$= 0 \tag{3-44}$$

A vector field $\mathbf{A}$ for which $\nabla \times \mathbf{A} = 0$ everywhere in space is said to be *irrotational.* If $\mathbf{A}$ is the gradient of a scalar function ϕ ($\mathbf{A} = \nabla\phi$), then it follows from Eq. (3-44) that $\mathbf{A}$ is irrotational. Thus, a *necessary* condition for a vector field to be represented as the gradient of a scalar field is that the vector field be irrotational. It can be shown through the use of Stokes' theorem (to be discussed in Sec. 3-12) that any irrotational vector field can be expressed as the gradient of a scalar field. Thus, the condition $\nabla \times \mathbf{A} = 0$ is both necessary and sufficient for the relation $\mathbf{A} = \nabla\phi$ to be valid. The scalar function ϕ is often called the *potential* of the vector function **A**.

3-6 Additional differential vector relationships

In applications of vector analysis it is often necessary to apply the vector differential operator ∇ to a product of functions or to operate several times in succession with ∇. In this section we list some useful identities which result from the definitions of the gradient, the divergence, and the curl.

Since the curl $\nabla \times \mathbf{A}$ of a vector field **A** is also a vector field, say **B**, one may take the curl of **B**:

$$\nabla \times \mathbf{B} = \nabla \times (\nabla \times \mathbf{A})$$

If we expand this expression in terms of the components of **A**, we obtain

$$\begin{aligned}\nabla \times (\nabla \times \mathbf{A}) &= \mathbf{i}\left(\frac{\partial^2 A_y}{\partial y\,\partial x} - \frac{\partial^2 A_x}{\partial y^2} - \frac{\partial^2 A_x}{\partial z^2} + \frac{\partial^2 A_z}{\partial z\,\partial x}\right) + \mathbf{j}(\ ) + \mathbf{k}(\ )\\ &= \mathbf{i}\left[\frac{\partial}{\partial x}\left(\frac{\partial A_y}{\partial y} + \frac{\partial A_z}{\partial z}\right) - \frac{\partial^2 A_x}{\partial y^2} - \frac{\partial^2 A_x}{\partial z^2}\right] + \mathbf{j}(\ ) + \mathbf{k}(\ )\\ &= \mathbf{i}\left[\frac{\partial}{\partial x}\left(\frac{\partial A_x}{\partial x} + \frac{\partial A_y}{\partial y} + \frac{\partial A_z}{\partial z}\right) - \left(\frac{\partial^2 A_x}{\partial x^2} + \frac{\partial^2 A_x}{\partial y^2} + \frac{\partial^2 A_x}{\partial z^2}\right)\right]\\ &\qquad + \mathbf{j}(\ ) + \mathbf{k}(\ )\\ &= \mathbf{i}\left[\frac{\partial}{\partial x}(\nabla \cdot \mathbf{A}) - \nabla^2 A_x\right] + \mathbf{j}(\ ) + \mathbf{k}(\ )\\ &= \nabla(\nabla \cdot \mathbf{A}) - \nabla^2\mathbf{A} \qquad (3\text{-}45)\end{aligned}$$

Similarly, by expanding the operator ∇ and the various vector functions in terms of their components, we may prove the following vector identities:

$$\nabla \cdot (\mathbf{A} \times \mathbf{B}) = \mathbf{B} \cdot (\nabla \times \mathbf{A}) - \mathbf{A} \cdot (\nabla \times \mathbf{B}) \qquad (3\text{-}46)$$

$$\nabla \times (\mathbf{A} \times \mathbf{B}) = \mathbf{A}(\nabla \cdot \mathbf{B}) - \mathbf{B}(\nabla \cdot \mathbf{A}) + (\mathbf{B} \cdot \nabla)\mathbf{A} - (\mathbf{A} \cdot \nabla)\mathbf{B} \qquad (3\text{-}47)$$

$$\nabla(\mathbf{A} \cdot \mathbf{B}) = (\mathbf{A} \cdot \nabla)\mathbf{B} + (\mathbf{B} \cdot \nabla)\mathbf{A} + \mathbf{A} \times (\nabla \times \mathbf{B}) + \mathbf{B} \times (\nabla \times \mathbf{A}) \qquad (3\text{-}48)$$

3-7 Line integral

Consider a vector field $\mathbf{A}(x,y,z)$ and a path c extending from the point P_1 to the

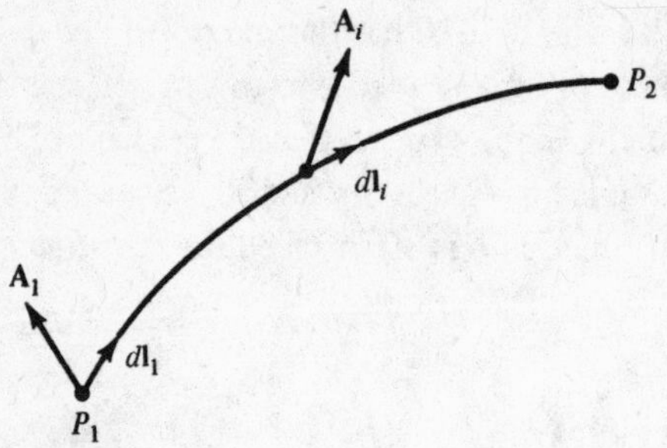

FIGURE 3-3

point P_2. Let the path c be divided into infinitesimal vector elements $d\mathbf{l}_1, d\mathbf{l}_2, \ldots, d\mathbf{l}_n$ as illustrated in Fig. 3-3. One can form the scalar products $\mathbf{A}_1 \cdot d\mathbf{l}_1, \mathbf{A}_2 \cdot d\mathbf{l}_2, \ldots, \mathbf{A}_n \cdot d\mathbf{l}_n$, where $\mathbf{A}_1, \mathbf{A}_2, \ldots, \mathbf{A}_n$ are the values of the vector function $\mathbf{A}$ at the locations of $d\mathbf{l}_1, d\mathbf{l}_2, \ldots, d\mathbf{l}_n$, respectively. The *line integral* of $\mathbf{A}$ over the path c is defined as

$$\lim_{n \to \infty} \sum_{i=1}^{n} \mathbf{A}_i \cdot d\mathbf{l}_i = \int_{P_1}^{P_2} \mathbf{A} \cdot d\mathbf{l} = -\int_{P_2}^{P_1} \mathbf{A} \cdot d\mathbf{l} \tag{3-49}$$

Thus the line integral of $\mathbf{A}$ along c is the integral of the component of $\mathbf{A}$ along the path.

In cartesian coordinates we may write $\mathbf{A} = \mathbf{i}A_x + \mathbf{j}A_y + \mathbf{k}A_z$ and $d\mathbf{l} = \mathbf{i}\,dx + \mathbf{j}\,dy + \mathbf{k}\,dz$, so that the line integral becomes

$$\int_c \mathbf{A} \cdot d\mathbf{l} = \int (A_x\,dx + A_y\,dy + A_z\,dz) \tag{3-50}$$

The value of the line integral of $\mathbf{A}$ between two points P_1 and P_2 may or may not depend on the specific path c which connects these two points. Suppose that $\mathbf{A}$ is the gradient of a scalar function $\phi\,(x,y,z)$,

$$\mathbf{A} = \nabla\phi \tag{3-51}$$

Then we have

$$\int_{P_1}^{P_2} \mathbf{A} \cdot d\mathbf{l} = \int_{P_1}^{P_2} (\nabla\phi) \cdot d\mathbf{l} = \int_{P_1}^{P_2} \left(\frac{\partial\phi}{\partial x}\,dx + \frac{\partial\phi}{\partial y}\,dy + \frac{\partial\phi}{\partial z}\,dz \right) \tag{3-52}$$

The total differential $d\phi$ of ϕ is just

$$d\phi = \frac{\partial\phi}{\partial x}\,dx + \frac{\partial\phi}{\partial y}\,dy + \frac{\partial\phi}{\partial z}\,dz \tag{3-53}$$

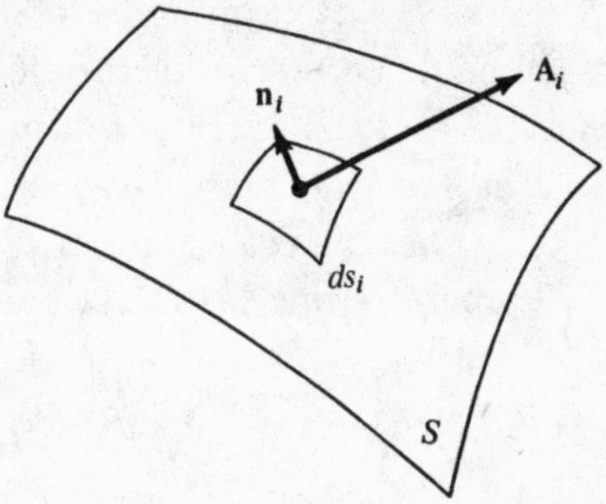

FIGURE 3-4

so that Eq. (3-52) becomes

$$\int_{P_1}^{P_2} \mathbf{A} \cdot d\mathbf{l} = \int_{P_1}^{P_2} d\phi = \phi(P_2) - \phi(P_1) \tag{3-54}$$

Thus, when $\mathbf{A} = \nabla\phi$ the line integral of $\mathbf{A}$ depends only on the end points and is independent of the path which connects them. The vector field $\mathbf{A}$ equals $\nabla\phi$ when $\mathbf{A}$ is irrotational ($\nabla \times \mathbf{A} = 0$), so that taking the curl of $\mathbf{A}$ may be used as a test for the validity of Eq. (3-54).

The line integral around a closed path (where the point P_2 coincides with P_1) is denoted by

$$\oint_C \mathbf{A} \cdot d\mathbf{l}$$

If $\mathbf{A} = \nabla\phi$, then $\phi(P_1)$ is equal to $\phi(P_2)$ since P_1 and P_2 are the same point. Equation (3-54) then yields

$$\oint \mathbf{A} \cdot d\mathbf{l} = 0 \tag{3-55}$$

for all closed paths.

3-8 Surface integral

Consider a vector field $\mathbf{A}(x,y,z)$ and a surface S. Divide the surface into n infinitesimal surface elements of areas $ds_1, \ldots, ds_n$. A typical surface element is shown in Fig. 3-4. Let $\mathbf{A}_1, \mathbf{A}_2, \ldots, \mathbf{A}_n$ be the values of $\mathbf{A}(x,y,z)$ in ds_1, $ds_2, \ldots, ds_n$, respectively.

We define $d\mathbf{s}_i$ as a vector of magnitude ds_i with a direction normal to the surface at the point in question. If the unit vector normal to the surface at that point is $\mathbf{n}_i$, then we have

$$d\mathbf{s}_i = \mathbf{n}_i \, ds_i \tag{3-56}$$

Since the surface elements are infinitesimally small, a $d\mathbf{s}_i$ is defined for each point on the surface S and, therefore, a vector field $d\mathbf{s}$ is determined. The direction of the unit vector $\mathbf{n}_i$ may be chosen in either of two possible directions, so that the sign of $\mathbf{n}_i$ is ambiguous. If the surface is closed, the *outward* normal is conventionally taken as positive.

The *surface integral* of the vector field $\mathbf{A}$ over the surface S is defined as

$$\lim_{n\to\infty} \sum_{i=1}^{n} \mathbf{A}_i \cdot d\mathbf{s}_i = \int_S \mathbf{A} \cdot d\mathbf{s} \qquad (3\text{-}57)$$

where the symbol $\int_S$ stands for a double integral over the surface S. The sign of the surface integral depends, of course, on which face of the surface is taken as positive. The vector function $d\mathbf{s}$ may be written in terms of its x, y, z components,

$$d\mathbf{s} = \mathbf{i}\, ds_x + \mathbf{j}\, ds_y + \mathbf{k}\, ds_z \qquad (3\text{-}58)$$

in which case Eq. (3-57) takes the form

$$\int_S \mathbf{A} \cdot d\mathbf{s} = \int_S (A_x\, ds_x + A_y\, ds_y + A_z\, ds_z) \qquad (3\text{-}59)$$

If the surface S is a closed surface, then the surface integral of $\mathbf{A}$ over S is denoted by

$$\oint_S \mathbf{A} \cdot d\mathbf{s}$$

and $d\mathbf{s}$ is directed outward.

3-9 Volume integral

If $\mathbf{A}(x,y,z)$ is a vector field and V is a region of three-dimensional space, the volume integral of $\mathbf{A}$ over V is

$$\int_V \mathbf{A}\, dv = \mathbf{i} \int_V A_x dv + \mathbf{j} \int_V A_y dv + \mathbf{k} \int_V A_z dv \qquad (3\text{-}60)$$

where dv is a volume element of V and $\int_V$ stands for a triple integral over the volume V. Thus, according to Eq. (3-60), the integration of $\mathbf{A}$ over V is carried out by performing three ordinary integrations involving only scalar functions.

3-10 Gauss' theorem

Gauss' theorem, sometimes called the *divergence theorem,* relates the surface integral of a vector function $\mathbf{A}$ to the volume integral of the divergence of $\mathbf{A}$. If S is the surface which encloses the volume V, then Gauss' theorem states that

$$\int_V (\nabla \cdot \mathbf{A})\, dv = \oint_S \mathbf{A} \cdot d\mathbf{s} \qquad (3\text{-}61)$$

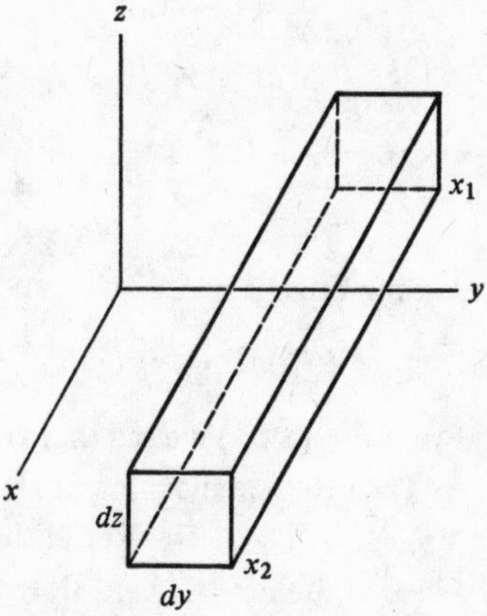

FIGURE 3-5

To prove Eq. (3-61), we first expand the left-hand side to give

$$\int_V (\nabla\cdot\mathbf{A})\,dv = \int_V \left(\frac{\partial A_x}{\partial x} + \frac{\partial A_y}{\partial y} + \frac{\partial A_z}{\partial z}\right) dx\,dy\,dz \tag{3-62}$$

Let us now consider the first integral on the right-hand side, namely,

$$\int_V \frac{\partial A_x}{\partial x}\,dx\,dy\,dz$$

We carry out the integration with respect to x along a rectangular tube of cross section $dy\,dz$ from $x = x_1$ to $x = x_2$ as shown in Fig. 3-5. Since y and z are held constant during this integration, the total derivative of A_x is simply $(\partial A_x/\partial x)dx$ and we obtain

$$\begin{aligned}\int_V \frac{\partial A_x}{\partial x}\,dx\,dy\,dz &= \oint_S dy\,dz \int_{x_1}^{x_2} dA_x \\ &= \oint_S [A_x(x_2,y,z) - A_x(x_1,y,z)]\,dy\,dz \\ &= \oint_S A_x(x_2,y,z)\,dy\,dz - \oint_S A_x(x_1,y,z)\,dy\,dz \end{aligned}\tag{3-63}$$

On the surface at x_2, the product $dy\,dz$ is just the x component ds_x of the vector ds. On the surface at x_1, the outward normal to the surface is in the negative x direction, so that $dy\,dz$ is equal to $-ds_x$. With this substitution Eq. (3-63) becomes

$$\int_V \frac{\partial A_x}{\partial x}\,dx\,dy\,dz = \oint_S A_x\,ds_x \tag{3-64}$$

By the same procedure we obtain

$$\int_V \frac{\partial A_y}{\partial y}\, dx\, dy\, dz = \oint_S A_y ds_y \tag{3-65}$$

$$\int_V \frac{\partial A_z}{\partial z}\, dx\, dy\, dz = \oint_S A_z ds_z \tag{3-66}$$

Adding Eqs. (3-64), (3-65), and (3-66), and referring to Eq. (3-62) we obtain

$$\int_V (\nabla \cdot \mathbf{A})\, dv = \oint_S (A_x ds_x + A_y ds_y + A_z ds_z) \tag{3-67}$$

From Eq. (3-59) we see that Gauss' theorem (3-61) has now been proved.

The physical meaning of Gauss' theorem may be illustrated by fluid flow as discussed in Sec. 3-4. Recall that if **D** is the rate of flow per unit area then $\nabla \cdot \mathbf{D}$ is the net outward flow rate per unit volume from a volume element dv. Therefore, the total outward rate of flow Q from a volume V is the integral of Eq. (3-38):

$$Q = \int_V dQ = \int_V \nabla \cdot \mathbf{D}\, dv \tag{3-68}$$

The total outward flow of fluid must, of course, be through the surface S of the volume V, so that

$$Q = \oint_S \mathbf{D} \cdot d\mathbf{s} \tag{3-69}$$

Comparison of Eqs. (3-68) and (3-69) leads directly to Gauss' theorem (3-61).

3-11 Green's theorem

Let the vector function **A** in Gauss' theorem (3-61) be of the form

$$\mathbf{A} = \phi \nabla \psi = \mathbf{i}\phi \frac{\partial \psi}{\partial x} + \mathbf{j}\phi \frac{\partial \psi}{\partial y} + \mathbf{k}\phi \frac{\partial \psi}{\partial z} \tag{3-70}$$

where ϕ and ψ are scalar fields. Then the divergence of **A** is

$$\begin{aligned} \nabla \cdot \mathbf{A} &= \frac{\partial}{\partial x}\left(\phi \frac{\partial \psi}{\partial x}\right) + \frac{\partial}{\partial y}\left(\phi \frac{\partial \psi}{\partial y}\right) + \frac{\partial}{\partial z}\left(\phi \frac{\partial \psi}{\partial z}\right) \\ &= \phi\left(\frac{\partial^2 \psi}{\partial x^2} + \frac{\partial^2 \psi}{\partial y^2} + \frac{\partial^2 \psi}{\partial z^2}\right) + \frac{\partial \phi}{\partial x}\frac{\partial \psi}{\partial x} + \frac{\partial \phi}{\partial y}\frac{\partial \psi}{\partial y} + \frac{\partial \phi}{\partial z}\frac{\partial \psi}{\partial z} \\ &= \phi \nabla^2 \psi + \nabla \phi \cdot \nabla \psi \end{aligned} \tag{3-71}$$

Substitution of this result into Eq. (3-61) yields

$$\int_V (\phi \nabla^2 \psi + \nabla \phi \cdot \nabla \psi)\, dv = \oint_S (\phi \nabla \psi) \cdot d\mathbf{s} \tag{3-72}$$

which is known as the *first form of Green's theorem.*

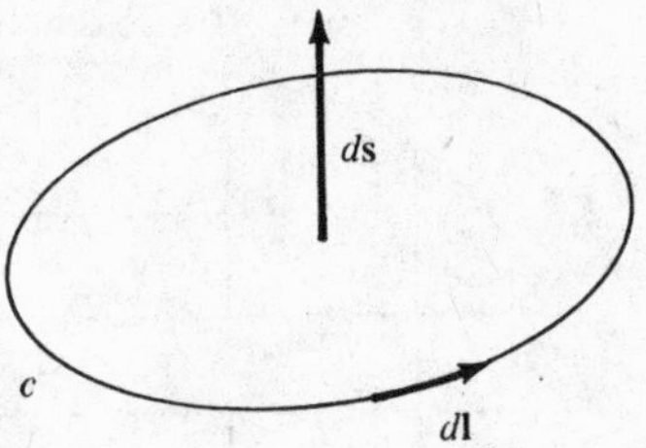

FIGURE 3-6

A *second form of Green's theorem* may be obtained as follows: First we exchange the roles of ϕ and ψ in Eq. (3-72) to obtain

$$\int_V (\psi \nabla^2 \phi + \nabla\phi \cdot \nabla\psi)\, dv = \oint_S (\psi \nabla\phi) \cdot d\mathbf{s} \tag{3-73}$$

We next subtract Eq. (3-73) from Eq. (3-72) and thereby obtain the desired result:

$$\int_V (\phi \nabla^2 \psi - \psi \nabla^2 \phi)\, dv = \oint_S (\phi \nabla\psi - \psi \nabla\phi) \cdot d\mathbf{s} \tag{3-74}$$

Another useful integral relationship may be obtained by setting $\psi = 1$ in Eq. (3-74). In this case, $\nabla\psi$ and $\nabla^2 \psi$ vanish and Eq. (3-74) reduces to

$$\int_V \nabla^2 \phi\, dv = \oint_S (\nabla\phi) \cdot d\mathbf{s} \tag{3-75}$$

3-12 Stokes' theorem

Stokes' theorem relates the line integral of a vector function **A** to the surface integral of the curl of **A**. If c is a closed path which bounds a surface S, then Stokes' theorem is

$$\oint_c \mathbf{A} \cdot d\mathbf{l} = \int_S (\nabla \times \mathbf{A}) \cdot d\mathbf{s} \tag{3-76}$$

It is assumed that c is a *simple* closed path, i.e., it does not cross itself.

Since the path c is usually taken as nonzero, the surface S is an open surface. In this case the positive direction for the vector $d\mathbf{s}$ is arbitrary as discussed in Sec. 3-8. We adopt here the convention that the positive direction for $d\mathbf{s}$ is determined by the right-hand rule: If the fingers of the right hand follow along the positive direction of the path c, then the thumb points in the positive direction of $d\mathbf{s}$ (see Fig. 3-6).

To prove Stokes' theorem (3-76), we first divide the surface S into n infinitesimal surface elements $d\mathbf{s}_1, d\mathbf{s}_2, \ldots, d\mathbf{s}_n$. These elements are sufficiently small that each can be regarded as a plane rectangle as shown in Fig. 3-7. We next calculate the sum of the line integrals of the vector field **A** along the contours bounding each surface element $d\mathbf{s}_1, d\mathbf{s}_2, \ldots, d\mathbf{s}_n$. The direction of the closed path around each surface element is taken in the same sense as the path c

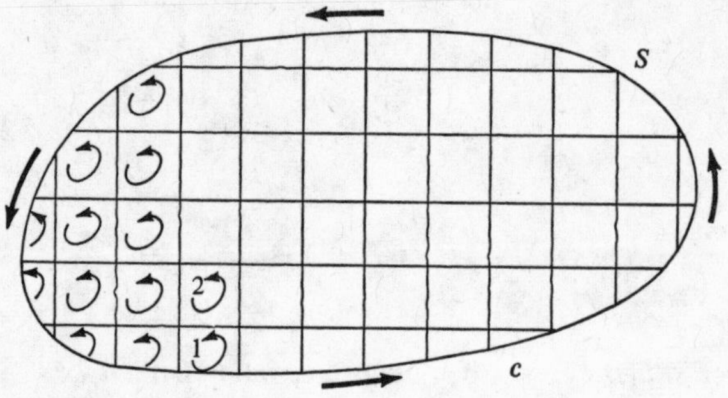

FIGURE 3-7

enclosing the surface S. These directions are indicated by the curved arrows in Fig. 3-7. We observe that elements 1 and 2 have a side in common. In the computation of the line integrals of elements 1 and 2, the common side is traced in opposite directions. Upon addition of the line integrals of the two adjacent elements, the contributions corresponding to the common side cancel. In this manner the contributions to the total line integral from the interior sides of the surface elements all cancel. The contributions from the exterior sides do not cancel, since they are traced only once in the summation, but yield the line integral over the path c. The net result is that the sum of the line integrals of $\mathbf{A}$ over the closed paths around the differential surface elements $ds_1, ds_2, \ldots, ds_n$ is just

$$\oint_c \mathbf{A} \cdot d\mathbf{l}$$

We now consider a typical rectangular planar surface element which lies in the yz plane and therefore has sides of length dy and dz as shown in Fig. 3-8. The corners of the rectangle are labeled P_1, P_2, P_3, P_4. Let A_y and A_z be the respective values of the y and z components of $\mathbf{A}(x,y,z)$ at the point P_1. The values of $A_y(x,y,z)$ and $A_z(x,y,z)$ at P_2, P_3, P_4 may be written as a Taylor series expansion about the point P_1. We retain in this expansion only the first two terms. Thus, the line integrals along the edges of the rectangular surface element in the direction shown in Fig. 3-8 are as follows:

Along P_1P_2: $\quad A_y\,dy$

Along P_2P_3: $\quad \left(A_z + \dfrac{\partial A_z}{\partial y}\,dy\right) dz$

Along P_3P_4: $\quad -\left(A_y + \dfrac{\partial A_y}{\partial z}\,dz\right) dy$

Along P_4P_1: $\quad -A_z\,dz$

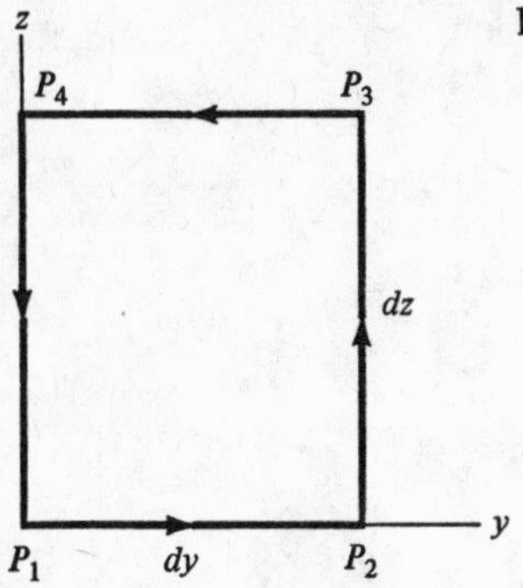

FIGURE 3-8

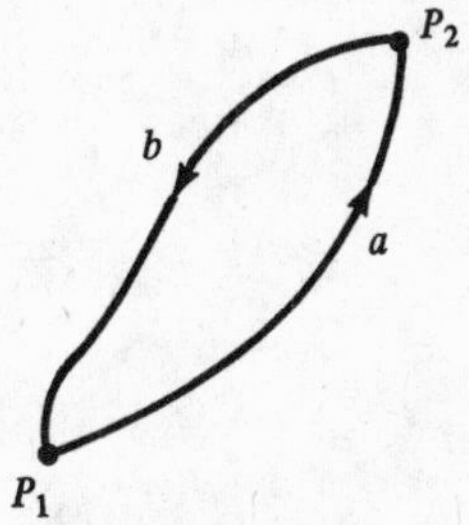

FIGURE 3-9

The last two expressions are negative because these portions of the path are traced in the negative y and z directions. Summing these contributions, we have

$$(\mathbf{A} \cdot d\mathbf{l})_{P_1P_2P_3P_4} = \left(\frac{\partial A_z}{\partial y} - \frac{\partial A_y}{\partial z}\right) dy\, dz \tag{3-77}$$

The right-hand side of Eq. (3-77) is the product of the x component of $\nabla \times \mathbf{A}$ and $dy\, dz = ds_x$, so that

$$(\mathbf{A} \cdot d\mathbf{l})_{P_1P_2P_3P_4} = (\nabla \times \mathbf{A})_x\, ds_x \tag{3-78}$$

We now sum the result expressed in Eq. (3-78) over all the surface elements $ds_1, ds_2, \ldots, ds_n$ and obtain

$$\sum_{i=1}^{n} (\mathbf{A} \cdot d\mathbf{l})_i = \sum_{i=1}^{n} [(\nabla \times \mathbf{A})_x\, ds_x + (\nabla \times \mathbf{A})_y\, ds_y + (\nabla \times \mathbf{A})_z\, ds_z] \tag{3-79}$$

The left-hand side has already been shown to yield

$$\oint \mathbf{A} \cdot d\mathbf{l}$$

In the limit as $n \to \infty$ the right-hand side becomes the surface integral

$$\int_S (\nabla \times \mathbf{A}) \cdot d\mathbf{s}$$

Thus, Eq. (3-76) is proved.

In Sec. 3-5 the statement was made that any irrotational vector field can be expressed as the gradient of a scalar field. This statement may be readily proved with Stokes' theorem. If $\nabla \times \mathbf{A} = 0$, then Stokes' theorem becomes

$$\oint_c \mathbf{A} \cdot d\mathbf{l} = 0 \tag{3-80}$$

Consider the line integral around an arbitrary closed path c as shown in Fig. 3-9.

The path c is divided into two parts a and b by the points P_1 and P_2. The line integral in Eq. (3-80) then becomes

$$\underset{a}{\int_{P_1}^{P_2}} \mathbf{A} \cdot d\mathbf{l} + \underset{b}{\int_{P_2}^{P_1}} \mathbf{A} \cdot d\mathbf{l} = 0 \tag{3-81}$$

so that

$$\underset{a}{\int_{P_1}^{P_2}} \mathbf{A} \cdot d\mathbf{l} = -\underset{b}{\int_{P_2}^{P_1}} \mathbf{A} \cdot d\mathbf{l} = \underset{b}{\int_{P_1}^{P_2}} \mathbf{A} \cdot d\mathbf{l} \tag{3-82}$$

Hence, for an irrotational vector field, the value of the line integral between two points is independent of the path of integration. This conclusion proves the existence of a function $\phi\,(x,y,z)$ with the property that

$$\int_{P_1}^{P_2} \mathbf{A} \cdot d\mathbf{l} = \phi(P_2) - \phi(P_1) \tag{3-83}$$

and, as shown in Sec. 3-7, **A** is then the gradient of $\phi\,(x,y,z)$.

PROBLEMS

1. The vectors $\mathbf{A}(t)$ and $\mathbf{B}(t)$ are given by

$$\mathbf{A} = t^2\mathbf{i} + \cos \omega t\,\mathbf{j} + \sin \omega t\,\mathbf{k}$$
$$\mathbf{B} = 4\mathbf{i} + 3t\mathbf{j} + (1 - t^2)\,\mathbf{k}$$

Evaluate at $t = 1$:

(*a*) $\dfrac{d\mathbf{A}}{dt}$ (*b*) $\dfrac{d^2\mathbf{A}}{dt^2}$

(*c*) $\left|\dfrac{d^2\mathbf{A}}{dt^2}\right|$ (*d*) $\dfrac{d\,(\mathbf{A} \cdot \mathbf{A})}{dt}$

(*e*) $\dfrac{d\,(\mathbf{A} \cdot \mathbf{B})}{dt}$ (*f*) $\dfrac{d\,(\mathbf{A} \times \mathbf{B})}{dt}$

(*g*) $\dfrac{d\,(\mathbf{A} + \mathbf{B})}{dt}$ (*h*) $\dfrac{d\,|\mathbf{A} + \mathbf{B}|}{dt}$

(*i*) $\dfrac{d}{dt}\left(\mathbf{A} \cdot \dfrac{d\mathbf{B}}{dt}\right)$

2. If $\phi = x \cos y + ze^y$, evaluate $\nabla\phi$ and $|\nabla\phi|$ at the point $x = 1, y = 2, z = 3$.

3. Determine the unit vectors normal to the surfaces $x^2 + y^2 + z^2 = 21$ and $xy + yz + xz = 14$ at the point $x = 2, y = 4, z = 1$. From these unit vectors, determine the angle between the two surfaces at this point.

4. If the vector **A** is given by

$$\mathbf{A} = \tfrac{1}{2}y^2 z\mathbf{i} + xyz\mathbf{j} + \tfrac{1}{2}xy^2\mathbf{k}$$

show that **A** is irrotational and find the scalar potential ϕ such that $\mathbf{A} = \nabla\phi$.

5. The vectors **A**, **B**, and **C** are given by

$$\mathbf{A} = \mathbf{i}x + \mathbf{j}y + \mathbf{k}z$$
$$\mathbf{B} = \mathbf{i}x^2 + \mathbf{j}y^2 + \mathbf{k}z^2$$
$$\mathbf{C} = \mathbf{i}z^2 + \mathbf{j}y + \mathbf{k}z$$

Find:

(*a*) $\nabla \cdot \mathbf{A}$ (*b*) $\nabla \cdot \mathbf{B}$
(*c*) $\nabla \cdot \mathbf{C}$ (*d*) $\nabla (\mathbf{A} \cdot \mathbf{B})$
(*e*) $\nabla^2 (\mathbf{B} \cdot \mathbf{A})$ (*f*) $\nabla \cdot [(\mathbf{A} \cdot \mathbf{B})\mathbf{C}]$
(*g*) $\nabla \times \mathbf{A}$ (*h*) $\nabla \times \mathbf{C}$
(*i*) $\nabla \times (\nabla \times \mathbf{C})$ (*j*) $\nabla \cdot (\nabla \times \mathbf{C})$
(*k*) $\nabla \cdot (\mathbf{A} \times \mathbf{C})$ (*l*) $\nabla \times (\mathbf{A} \times \mathbf{C})$

6. Prove the relationships in Eqs. (3-46) to (3-48).

7. The vector **A** is given by

$$\mathbf{A} = \frac{1}{x^2}\mathbf{i} - \frac{2}{y^2}\mathbf{j}$$

Evaluate $\int \mathbf{A} \cdot d\mathbf{l}$ from the point (1,2) to the point (2,3).

8. A vector **A** is given by

$$\mathbf{A} = \mathbf{i}xy^2 + \mathbf{j}y^3 + \mathbf{k}x^2e^y$$

(*a*) Evaluate the volume integral $\int_V \nabla \cdot \mathbf{A}\, dv$ throughout the cube bounded by the planes $x = 0, x = a, y = 0, y = a, z = 0, z = a$.

(*b*) Evaluate the surface integral $\oint_S \mathbf{A} \cdot d\mathbf{s}$ over the surface of the cube and verify Gauss' theorem for this example.

9. A vector field **A** is given by

$$\mathbf{A} = -3x^2y\mathbf{i} + (x^3 + y^3)\mathbf{j}$$

(a) Evaluate the surface integral $\int_S (\nabla \times ds)$ over a rectangular surface in the $z = 0$ plane bounded by the lines $x = 0, x = a, y = 0, y = b$.

(b) Evaluate the line integral $\oint \mathbf{A} \cdot d\mathbf{l}$ over the rectangular path and verify Stokes' theorem for this example.

10. Evaluate the line integral $\oint \mathbf{A} \cdot d\mathbf{l}$, where

$$\mathbf{A} = \mathbf{i}x^2 + \mathbf{j}z^2 + \mathbf{k}y^2$$

and the path of integration is around the ellipse $(x/a)^2 + (y/b)^2 = 1$.

11. A vector **A** is given by

$$\mathbf{A} = \mathbf{i}(x - z) + \mathbf{j}(x + 2y - 3z) + \mathbf{k}(3x^2 - 5y)$$

Evaluate the line integral $\oint \mathbf{A} \cdot d\mathbf{l}$ over the circular path $x^2 + y^2 = a^2$.

12. If the vector **A** is given by

$$\mathbf{A} = \mathbf{i}\cos y + \mathbf{j}x(y^2 - \sin y) + \mathbf{k}\sin y \cos y$$

use Stokes' theorem to evaluate the line integral $\oint \mathbf{A} \cdot d\mathbf{l}$ over the closed path in the $z = 0$ plane bounded by the x axis, the line $x = 2$, and the line $x = y$.

Chapter four

Electromagnetic Theory

4-1 Introduction

The theory of electromagnetism provides an excellent example of the application of vector calculus to physical phenomena. Moreover, a knowledge of electromagnetic theory is becoming increasingly more important in the study of chemistry. For these reasons, this chapter is devoted to an introduction to electromagnetic theory. The discussion here is too short to constitute anything but the minimum exposure to the subject. The reader who wishes to gain a firm working knowledge of electromagnetic theory should consult one or more of the many excellent books on the subject.

The assumption is made here that the reader has been exposed to an introduction to electricity and magnetism at the level of first-year college physics. Such an introduction usually involves discussions of experiments that lead to the concepts of electric and magnetic fields. In our treatment here we present a deductive approach, in which the existence of electric and magnetic fields and the equations which govern them are postulated. The interpretation of experimental results is then deduced by the application of these postulates. The acceptance of these postulates comes from the agreement, without exception, of deduction with observation for many experiments.

The emphasis in the discussion of electromagnetic theory presented here is on the mathematical development rather than on physical phenomena. Moreover, those aspects involving vector analysis are stressed.

4-2 Units

The study of electromagnetic theory is complicated by the existence of several systems of units and dimensions for expressing electromagnetic quantities. As a result, the mathematical equations relating various electromagnetic quantities take different forms according to the system of units being employed.

Although there are many such systems of units, only two are now in common use: The gaussian or CGS (centimeter-gram-second) system and the SI or rationalized MKSA (meter-kilogram-second-ampere) system. SI units (Système Internationale d'Unités) are being adopted worldwide as the only form of the metric system of units to be used. However, the gaussian system is more convenient for applications to the molecular phenomena studied in chemistry, so that much of the existing literature employs these units. Consequently, we present the electromagnetic equations in this chapter in both CGS and SI units.

By way of illustration, we consider Coulomb's law in the two systems of units,

$$\text{(CGS)} \qquad |\mathbf{f}| = \frac{q_1 q_2}{r^2}$$

$$\text{(SI)} \qquad |\mathbf{f}| = \frac{q_1 q_2}{4\pi\epsilon_0 r^2}$$

where **f** is the force between two electric charges q_1 and q_2 separated by a distance r in a vacuum (free space). In CGS units, **f** is in dynes, r in centimeters, and q in electrostatic units (esu). In SI units, **f** is in newtons, r in meters, and q in coulombs. The quantity ϵ_0 is called the *permittivity of free space* and has the numerical value $\epsilon_0 = 10^7/4\pi c^2 = 8.854 \times 10^{-12}$ F/m ($C^2\ N^{-1}\ m^{-2}$), where c is the velocity of light in a vacuum (2.998×10^8 m/s). This particular numerical value for ϵ_0 is the result of the definition of the ampere as a basic unit and hence of the coulomb as a derived unit.

More detailed discussions on units in electromagnetic theory may be found in most textbooks on the subject and in the literature.[1]

4-3 Charges and currents

An *electromagnetic field* is characterized by the existence of four vector fields: the electric field intensity **E**, the magnetic field intensity **H**, the electric displacement **D**, and the magnetic induction **B**. These vector fields and their space and time derivatives are finite and are continuous functions of position and time at every point in whose neighborhood the physical properties of the medium are continuous. Discontinuities in **E**, **H**, **D**, and **B** occur on surfaces

[1] S. Chapman, *Am. J. Phys.*, **24**: 162 (1956); N. H. Davies, *Chem. Brit.*, **7**: 331 (1971); C. H. Page, *Am. J. Phys.*, **38**: 421 (1970); D. H. Smith, *Contemp. Phys.*, **11**: 287 (1970).

where physical properties are also discontinuous, for example, on surfaces between various phases in a material system.

The sources of an electromagnetic field are distributions of electric charge and of electric current. These distributions are assumed to be continuous and are specified by the *density of charge* ρ (a scalar field) and the *electric current density* **J** (a vector field).

The charge density $\rho\,(x,y,z,t)$ is a scalar field which gives the electric charge per unit volume at time t in the neighborhood of the point x,y,z. The function may take on positive or negative values. The volume integral of ρ is the total net charge Q within the volume V at time t:

$$Q = \int_V \rho(x,y,z,t)\, dv \tag{4-1}$$

The ordered motion of electric charge is an electric current, which is specified by the current density **J**. The vector field $\mathbf{J}\,(x,y,z,t)$ is the charge which in unit time crosses a unit area of surface perpendicular to the direction of flow at the point x,y,z and at time t. The electric current **I** across any arbitrary surface S is simply the surface integral of the current density:

$$\mathbf{I} = \int_S \mathbf{J} \cdot d\mathbf{s} \tag{4-2}$$

We now consider an arbitrary *closed* surface S. The current flowing through the surface is given by Eq. (4-2). Thus, **I** is the rate of loss of positive charge from the volume V which S encloses. (If negative charges are involved, then **I** is the rate of addition of negative charge to V.) If we assume that electric charge is conserved, then the rate of loss of positive charge due to the current **I** through S must equal the rate of change of the charge in V, so that

$$\oint_S \mathbf{J} \cdot d\mathbf{s} = -\frac{\partial}{\partial t}\int_V \rho\, dv \tag{4-3}$$

A minus sign appears on the right-hand side of Eq. (4-3) because positive charge is lost from V. Since the volume V is fixed in space and does not change with time, the order of differentiation and integration may be interchanged to give

$$\oint_S \mathbf{J} \cdot d\mathbf{s} = -\int_V \frac{\partial \rho}{\partial t}\, dv \tag{4-4}$$

We next apply Gauss' theorem (3-61) to the left-hand side and thereby obtain

$$\int_V \left(\nabla \cdot \mathbf{J} + \frac{\partial \rho}{\partial t}\right) dv = 0 \tag{4-5}$$

Since the volume V under consideration is arbitrary, the only way for Eq. (4-5) to be valid for all volumes V is for the integrand to vanish. We therefore obtain the *equation of continuity*

$$\nabla \cdot \mathbf{J} + \frac{\partial \rho}{\partial t} = 0 \tag{4-6}$$

expressing the conservation of charge in the neighborhood of each point in space.

If the charge density ρ does not change with time, Eq. (4-6) reduces to

$$\nabla \cdot \mathbf{J} = 0 \tag{4-7}$$

and **J** is solenoidal.

An electric field exerts a coulombic force on electric charges, and a magnetic field exerts a force on electric currents. The force field **F** (x,y,z,t) is given by

$$\begin{aligned} &\text{(CGS)} \qquad \mathbf{F} = \rho\mathbf{E} + \frac{1}{c}(\mathbf{J} \times \mathbf{B}) \\ &\text{(SI)} \qquad \mathbf{F} = \rho\mathbf{E} + (\mathbf{J} \times \mathbf{B}) \end{aligned} \tag{4-8}$$

where the first term on the right-hand side is the coulombic force and the second term is the magnetic force. The field **F** is the force per unit volume (force density). The force **f** exerted on a single particle of charge e and velocity **v** is

$$\begin{aligned} &\text{(CGS)} \qquad \mathbf{f} = e\mathbf{E} + \frac{e}{c}(\mathbf{v} \times \mathbf{B}) \\ &\text{(SI)} \qquad \mathbf{f} = e\mathbf{E} + e(\mathbf{v} \times \mathbf{B}) \end{aligned} \tag{4-9}$$

since the current for such a particle is e**v**.

4-4 Maxwell's equations

The electromagnetic field vectors **E, H, D,** and **B** are related to each other and to the charge density ρ and the current density **J** by *Maxwell's equations*:

(CGS)	(SI)	
$\nabla \times \mathbf{E} + \frac{1}{c}\frac{\partial \mathbf{B}}{\partial t} = 0$	$\nabla \times \mathbf{E} + \frac{\partial \mathbf{B}}{\partial t} = 0$	(4-10)
$\nabla \times \mathbf{H} - \frac{1}{c}\frac{\partial \mathbf{D}}{\partial t} = \frac{4\pi}{c}\mathbf{J}$	$\nabla \times \mathbf{H} - \frac{\partial \mathbf{D}}{\partial t} = \mathbf{J}$	(4-11)
$\nabla \cdot \mathbf{D} = 4\pi\rho$	$\nabla \cdot \mathbf{D} = \rho$	(4-12)
$\nabla \cdot \mathbf{B} = 0$	$\nabla \cdot \mathbf{B} = 0$	(4-13)

where, as before, c is the velocity of light in vacuum (2.998×10^{10} cm/s =

2.998×10^8 m/s). Maxwell's equations are mathematical expressions in differential form for physical laws deduced from experiment. In the remainder of this section we consider these relationships one by one.

The Maxwell equation (4-10) leads directly to Faraday's induction law. To show this relationship we integrate the normal components of the vectors in Eq. (4-10) over any regular surface S bounded by a closed contour c:

$$\text{(CGS)} \qquad \int_S (\nabla \times \mathbf{E}) \cdot d\mathbf{s} = -\frac{1}{c}\int_S \frac{\partial \mathbf{B}}{\partial t} \cdot d\mathbf{s}$$

$$\text{(SI)} \qquad \int_S (\nabla \times \mathbf{E}) \cdot d\mathbf{s} = -\int_S \frac{\partial \mathbf{B}}{\partial t} \cdot d\mathbf{s} \tag{4-14}$$

Application of Stokes' theorem (3-76) to the left-hand side yields

$$\int_S (\nabla \times \mathbf{E}) \cdot d\mathbf{s} = \oint_c \mathbf{E} \cdot d\mathbf{l} \tag{4-15}$$

The quantity $\mathbf{E} \cdot d\mathbf{l}$ is the scalar product of a distance $d\mathbf{l}$ and a force $\mathbf{E}$ on a unit charge. Hence, $\mathbf{E} \cdot d\mathbf{l}$ is the work done in moving the unit charge a distance $d\mathbf{l}$. The integral $\oint_c \mathbf{E} \cdot d\mathbf{l}$ is the work done in moving the unit charge around the contour c. This work is just the electromotive force (emf) or potential difference in the loop c and is given the symbol ϕ. Since the surface S is fixed and does not change with time, the integral on the right-hand side of Eq. (4-14) may be written

$$\int_S \frac{\partial \mathbf{B}}{\partial t} \cdot d\mathbf{s} = \frac{\partial}{\partial t} \int_S \mathbf{B} \cdot d\mathbf{s} \tag{4-16}$$

By definition the integral $\int_S \mathbf{B} \cdot d\mathbf{s}$ is the flux or flow of $\mathbf{B}$ through the surface S. This integral is called the *magnetic flux* and is given the symbol Φ. Thus, Eq. (4-14) takes the form of Faraday's induction law,

$$\text{(CGS)} \qquad \phi = -\frac{1}{c}\frac{\partial \Phi}{\partial t}$$

$$\text{(SI)} \qquad \phi = -\frac{\partial \Phi}{\partial t} \tag{4-17}$$

which states that in a closed circuit the potential generated equals the rate of change of magnetic flux.

The second Maxwell equation (4-11) leads to Ampere's law. To demonstrate this relationship, we take the surface integral of each side of Eq. (4-11):

$$\text{(CGS)} \qquad \int_S (\nabla \times \mathbf{H}) \cdot d\mathbf{s} - \frac{1}{c}\int_S \frac{\partial \mathbf{D}}{\partial t} \cdot d\mathbf{s} = \frac{4\pi}{c}\int_S \mathbf{J} \cdot d\mathbf{s}$$

$$\text{(SI)} \qquad \int_S (\nabla \times \mathbf{H}) \cdot d\mathbf{s} - \int_S \frac{\partial \mathbf{D}}{\partial t} \cdot d\mathbf{s} = \int_S \mathbf{J} \cdot d\mathbf{s} \tag{4-18}$$

For the steady-state situation, the time derivative of **D** vanishes. Moreover, the integral on the right-hand side is just the electric current **I**. We may also apply Stokes' theorem (3-76) to the left-hand side. Thus, Eq. (4-18) becomes Ampere's law,

$$\begin{aligned} &\text{(CGS)} \qquad \int_C \mathbf{H} \cdot d\mathbf{l} = \frac{4\pi}{c}\mathbf{I} \\ &\text{(SI)} \qquad \oint_C \mathbf{H} \cdot d\mathbf{l} = \mathbf{I} \end{aligned} \tag{4-19}$$

which states that the line integral of the magnetic field around any closed loop equals the electric current flowing through the surface enclosed by the loop.

If we take the divergence of both sides of Eq. (4-11) and apply the result to a system which is not in a steady state, we have

$$\begin{aligned} &\text{(CGS)} \qquad \nabla \cdot \nabla \times \mathbf{H} = \frac{1}{c}\frac{\partial \nabla \cdot \mathbf{D}}{\partial t} + \frac{4\pi}{c}\nabla \cdot \mathbf{J} \\ &\text{(SI)} \qquad \nabla \cdot \nabla \times \mathbf{H} = \frac{\partial \nabla \cdot \mathbf{D}}{\partial t} + \nabla \cdot \mathbf{J} \end{aligned} \tag{4-20}$$

where the order of differentiation of **D** has been changed. The left-hand side vanishes by virtue of Eq. (3-43). If we use the Maxwell equation (4-12) to replace $\nabla \cdot \mathbf{D}$, Eq. (4-20) becomes

$$\frac{\partial \rho}{\partial t} + \nabla \cdot \mathbf{J} = 0 \tag{4-21}$$

which is the equation of continuity (4-6).

If we take the volume integral of the third Maxwell equation (4-12), we obtain

$$\begin{aligned} &\text{(CGS)} \qquad \int_V (\nabla \cdot \mathbf{D})\, dv = 4\pi \int_V \rho\, dv \\ &\text{(SI)} \qquad \int_V (\nabla \cdot \mathbf{D})\, dv = \int_V \rho\, dv \end{aligned} \tag{4-22}$$

The integral on the right-hand side is the total charge Q_{in} inside the volume V. The left-hand side can be transformed to a surface integral by means of the divergence theorem (3-61). Equations (4-22) then become

$$\begin{aligned} &\text{(CGS)} \qquad \oint_S \mathbf{D} \cdot d\mathbf{s} = 4\pi Q_{in} \\ &\text{(SI)} \qquad \oint_S \mathbf{D} \cdot d\mathbf{s} = Q_{in} \end{aligned} \tag{4-23}$$

Equation (4-23) is Gauss' law, which states that the flux of **D** through a closed surface S equals the charge inside the volume V enclosed by S. The derivation of Gauss' law (4-23) from electrostatics depends on the use of a central inverse-square law for the force between charges. Hence, the Maxwell equation (4-12) is often said to be a statement in differential form of Coulomb's law.

The physical interpretation of the fourth Maxwell equation (4-13) is obtained by taking its volume integral and applying Gauss' theorem (3-61):

$$\int_V (\nabla \cdot \mathbf{B})\, dv = \oint_S \mathbf{B} \cdot d\mathbf{s} = 0 \tag{4-24}$$

This expression states that the total flux of **B** crossing any closed regular surface is zero. Hence, there are no free magnetic poles. Thus, although point electric charges – both positive and negative – exist, isolated magnetic poles do not. All magnetic fields are caused by electric currents.

4-5 Material equations

Maxwell's equations connect the four basic vector fields **E**, **H**, **D**, and **B**. However, these four partial differential equations are not sufficient to determine uniquely these vector fields from a given distribution of electric charges and currents. It is necessary to supplement Maxwell's equations by *constitutive relations* which describe the behavior of substances under the influence of an electromagnetic field.

If the physical properties of a body in the neighborhood of each point within the body are the same in all directions, the body is *isotropic.* For an isotropic homogeneous body at rest or in slow motion relative to the electromagnetic field, the *constitutive relations* or *material equations* are

$$\mathbf{D} = \epsilon\epsilon_0 \mathbf{E} \tag{4-25}$$

$$\mathbf{B} = \mu\mu_0 \mathbf{H} \tag{4-26}$$

$$\mathbf{J} = \sigma \mathbf{E} \tag{4-27}$$

Equations (4-25) to (4-27) are written in a form which is applicable in both CGS and SI units: ϵ is the *relative permittivity,* μ the *relative permeability,* σ the *conductivity* of the medium; ϵ_0 is the *permittivity of free space* ($\epsilon_0 = 8.854 \times 10^{-12}$ F/m in SI units and equals unity without dimensions in CGS units; see Sec. 4-2), and μ_0 is the *permeability of free space* ($\mu_0 = 4\pi \times 10^{-7}$ H/m in SI units and equals unity without dimensions in CGS units). When Eqs. (4-25) and (4-26) are expressed in CGS units, the factors ϵ_0, μ_0 are usually omitted. Equation (4-27) is the differential form of Ohm's law.

The factors ϵ, μ, and σ are always positive, so that **D** and **J** are parallel to **E** and **B** is parallel to **H**. The relative permittivity ϵ, often called the *dielectric constant,* is a dimensionless quantity and has the same value in both CGS and SI units. Its value in a medium is greater than unity and can cover a wide range of values (1.000264 for gaseous hydrogen at 0°C, 81.1 for liquid water at 20°C). The relative permeability μ is also dimensionless, with the same value for both sets of units. For most substances μ is unity or nearly so. However, for *magnetic* substances μ is not unity. If μ is less than unity, the substance is *diamagnetic* (e.g., graphite, hydrogen, water), whereas if μ is greater than unity, the substance is *paramagnetic* (e.g., chromium, oxygen, nitrogen). Substances for which the conductivity is negligibly small ($\sigma \approx 0$) are called *insulators* or *dielectrics.* Otherwise the substance is a *conductor.* The range of values for σ is very large ($\approx 10^{-17}$ mho/m for fused quartz, 6.1×10^7 mho/m for silver). In a vacuum or free space, ϵ and μ are both unity and σ is zero.

The constitutive relations (4-25) to (4-27) assume that ϵ, μ, and σ are constants for a given isotropic homogeneous substance and are independent of the field strengths. The relation (4-26) is not valid for *ferromagnetic* substances (e.g., iron, cobalt, nickel) because of their high degree of magnetization ($\mu >> 1$). For ferromagnetic substances **B** depends on the magnetic history of the material as well as on the instantaneous value of **H**. Similarly, in rapidly alternating electric fields **E**, the relative permittivity ϵ is a function of the frequency of the field and is not constant. In nonhomogeneous material, ϵ, μ, and σ are not constants but become functions of x, y, z. For anisotropic substances Eqs. (4-25) to (4-27) are not valid. Equation (4-27), for example, is replaced by

$$\begin{aligned} J_x &= \sigma_{11}E_x + \sigma_{12}E_y + \sigma_{13}E_z \\ J_y &= \sigma_{21}E_x + \sigma_{22}E_y + \sigma_{23}E_z \\ J_z &= \sigma_{31}E_x + \sigma_{32}E_y + \sigma_{33}E_z \end{aligned} \tag{4-28}$$

with analogous equations replacing Eqs. (4-25) and (4-26). The coefficients σ_{ij} are components of a symmetric tensor or dyadic (see Chap. 5).

It is sometimes convenient to introduce two new vector fields, the *electric polarization* **P** and the *magnetic polarization* **M**, by means of the relations

$$\begin{aligned} &\text{(CGS)} \qquad \mathbf{D} = \mathbf{E} + 4\pi\mathbf{P} \qquad \mathbf{B} = \mathbf{H} + 4\pi\mathbf{M} \\ &\text{(SI)} \qquad \mathbf{D} = \epsilon_0\mathbf{E} + \mathbf{P} \qquad \mathbf{B} = \mu_0(\mathbf{H} + \mathbf{M}) \end{aligned} \tag{4-29}$$

Both **P** and **M** represent the influence of matter on the field vectors and vanish in free space. In isotropic homogeneous media Eqs. (4-25), (4-26), and (4-29) give

$$\begin{aligned} &\text{(CGS)} \qquad \mathbf{P} = \frac{\epsilon - 1}{4\pi}\mathbf{E} = \chi_e\mathbf{E} \qquad \mathbf{M} = \frac{\mu - 1}{4\pi}\mathbf{H} = \chi_m\mathbf{H} \\ &\text{(SI)} \qquad \mathbf{P} = (\epsilon - 1)\epsilon_0\mathbf{E} = \chi_e\epsilon_0\mathbf{E} \qquad \mathbf{M} = (\mu - 1)\mathbf{H} = \chi_m\mathbf{H} \end{aligned} \tag{4-30}$$

where χ_e is the *electric susceptibility* and χ_m is the *magnetic susceptibility.* These quantities are defined by Eqs. (4-30).

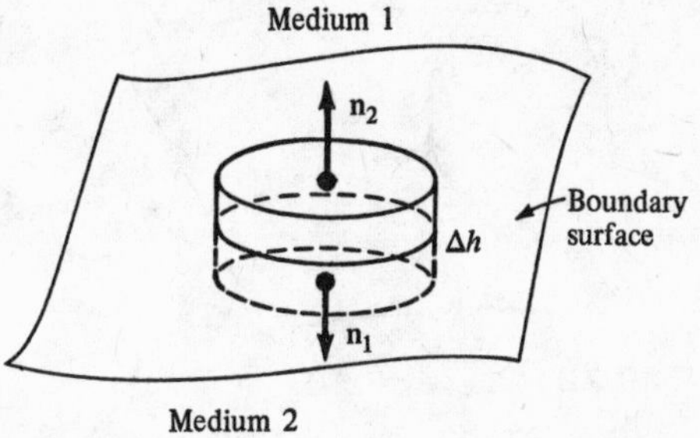

FIGURE 4-1

4-6 Boundary conditions at a surface of discontinuity

Maxwell's equations relate the electromagnetic vector fields at points in whose neighborhood the physical properties of the medium are invariant or vary continuously. However, on a boundary surface between two media, discontinuous changes occur in ϵ, μ, and σ and hence, in **E**, **H**, **D**, and **B**.

Consider a boundary surface which separates medium 1 from medium 2. Within the two media the vector fields and their first derivatives are continuous functions of position and time. Through the boundary surface we now construct a cylindrical volume element or "pillbox" as shown in Fig. 4-1. The axis of the cylinder is normal to the surface. One end lies in region 1 and the other in region 2. The height of the cylindrical volume element is Δh and the area of each end is Δs.

We first consider the vector field **D**, which is governed by Eq. (4-12). The volume integral of Eq. (4-12) over the cylindrical volume element is

$$\text{(CGS)} \qquad \int_V \nabla \cdot \mathbf{D}\, dv = 4\pi \int_V \rho\, dv$$
$$\text{(SI)} \qquad \int_V \nabla \cdot \mathbf{D}\, dv = \int_V \rho\, dv \tag{4-31}$$

Application of Gauss' theorem to the left-hand side and application of Eq. (4-1) to the right-hand side yield

$$\text{(CGS)} \qquad \oint_S \mathbf{D} \cdot d\mathbf{s} = 4\pi Q$$
$$\text{(SI)} \qquad \oint_S \mathbf{D} \cdot d\mathbf{s} = Q \tag{4-32}$$

where S is the surface of the cylinder and Q is the charge inside. The contribution of the walls to the surface integral is proportional to the height Δh. In the limit as $\Delta h \to 0$, the ends of the cylinder lie just on either side of the boundary surface and the contribution to the surface integral of the walls is vanishingly small. We assume that the area Δs of each end of the cylinder is sufficiently small that **D** has a constant value over the entire end. The value of **D** on the end in medium 1 is denoted

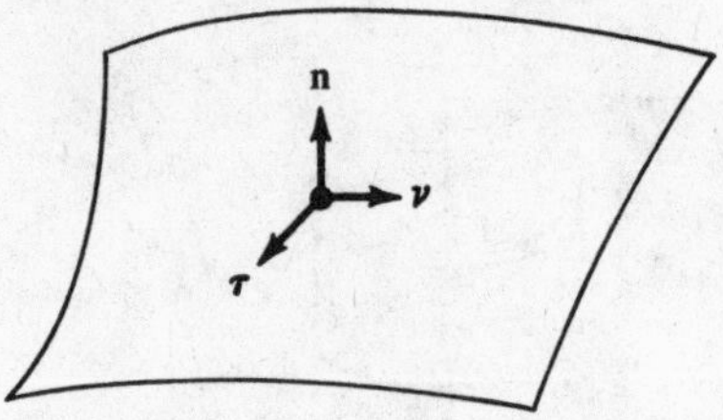

FIGURE 4-2

by $\mathbf{D}_1$ and that on the end in medium 2 by $\mathbf{D}_2$. If $\mathbf{n}_1$ and $\mathbf{n}_2$ are the outward normals to the ends (see Fig. 4-1), then the surface integral in Eqs. (4-32) in the limit $\Delta h \to 0$ becomes

$$\oint_S \mathbf{D} \cdot d\mathbf{s} = (\mathbf{D}_1 \cdot \mathbf{n}_1 + \mathbf{D}_2 \cdot \mathbf{n}_2)\,\Delta s \tag{4-33}$$

If we construct the unit vector $\mathbf{n}$ normal to the boundary surface and directed from medium 1 to medium 2, we have $\mathbf{n}_1 = -\mathbf{n}$ and $\mathbf{n}_2 = \mathbf{n}$, and Eq. (4-33) is

$$\oint_S \mathbf{D} \cdot d\mathbf{s} = (\mathbf{D}_2 - \mathbf{D}_1) \cdot \mathbf{n}\,\Delta s \tag{4-34}$$

In the limit as $\Delta h \to 0$, the right-hand side of Eq. (4-32) must also be rewritten. Since the conservation of charge requires that Q remain constant as the cylindrical volume element shrinks, the charge distribution reduces to a surface density of charge ω. Thus, the charge Q within the volume element becomes

$$Q = \omega\,\Delta s \tag{4-35}$$

Combining Eqs. (4-32), (4-34), and (4-35), we obtain

$$\begin{aligned} &\text{(CGS)} \quad (\mathbf{D}_2 - \mathbf{D}_1) \cdot \mathbf{n} = 4\pi\omega \\ &\text{(SI)} \quad\;\; (\mathbf{D}_2 - \mathbf{D}_1) \cdot \mathbf{n} = \omega \end{aligned} \tag{4-36}$$

Let $\boldsymbol{\tau}$ and $\boldsymbol{\nu}$ be mutually perpendicular unit vectors which are tangential to the boundary surface between media 1 and 2 (see Fig. 4-2). Then $\mathbf{n}, \boldsymbol{\tau}, \boldsymbol{\nu}$ form a basis set of orthogonal unit vectors and $\mathbf{D}$ can be written in terms of its three components in the $\mathbf{n}$, $\boldsymbol{\tau}$, and $\boldsymbol{\nu}$ directions:

$$\mathbf{D} = \mathbf{n}D_n + \boldsymbol{\tau}D_\tau + \boldsymbol{\nu}D_\nu \tag{4-37}$$

Substitution of Eq. (4-37) into Eqs. (4-36) gives

$$\begin{aligned} &\text{(CGS)} \quad D_{n2} - D_{n1} = 4\pi\omega \\ &\text{(SI)} \quad\;\; D_{n2} - D_{n1} = \omega \end{aligned} \tag{4-38}$$

Thus, the normal component of the electric displacement vector is discontinuous on the boundary surface by an amount related to the surface density of charge. If there is no surface charge ($\omega = 0$), the normal component of $\mathbf{D}$ is continuous

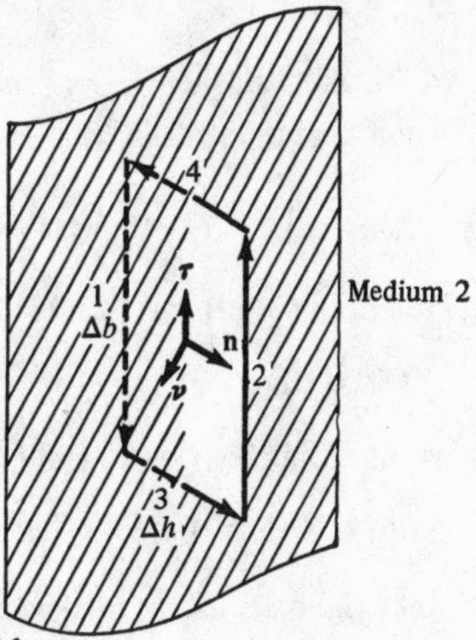

FIGURE 4-3

across the boundary.

From the relation (4-13) a similar argument yields

$$(\mathbf{B}_2 - \mathbf{B}_1) \cdot \mathbf{n} = 0 \tag{4-39}$$

so that the normal component of **B** is continuous across any surface of discontinuity.

To discuss the boundary conditions for **E** and **H** we consider a rectangular surface element S which is perpendicular to the boundary surface between the two media (see Fig. 4-3). The side in medium 1 is referred to as side 1, in medium 2 as side 2, and the ends as sides 3 and 4. The sides 1 and 2 are of length Δb, the sides 3 and 4 of length Δh. The unit vector **n** is normal to the boundary surface and directed from region 1 to region 2. The unit vectors $\boldsymbol{\tau}$ and $\boldsymbol{\nu}$ lie in the plane of the boundary surface with $\boldsymbol{\tau}$ parallel to sides 1 and 2 and $\boldsymbol{\nu}$ parallel to sides 3 and 4. The three unit vectors form the basis of a right-handed coordinate system according to the relation

$$\boldsymbol{\tau} = \boldsymbol{\nu} \times \mathbf{n} \tag{4-40}$$

The electric field **E** is governed by Eq. (4-10). Taking the surface integral of Eq. (4-10) over the rectangular surface element S, we have

$$\text{(CGS)} \quad \int_S (\nabla \times \mathbf{E}) \cdot d\mathbf{s} + \frac{1}{c}\int_S \frac{\partial \mathbf{B}}{\partial t} \cdot d\mathbf{s} = \oint_c \mathbf{E} \cdot d\mathbf{l} + \frac{1}{c}\int_S \frac{\partial \mathbf{B}}{\partial t} \cdot d\mathbf{s} = 0$$

$$\text{(SI)} \quad \int_S (\nabla \times \mathbf{E}) \cdot d\mathbf{s} + \int_S \frac{\partial \mathbf{B}}{\partial t} \cdot d\mathbf{s} = \oint_c \mathbf{E} \cdot d\mathbf{l} + \int_S \frac{\partial \mathbf{B}}{\partial t} \cdot d\mathbf{s} = 0 \tag{4-41}$$

where Stokes' theorem has been applied and c is the contour formed by sides 1, 3, 2, 4 with the positive direction as shown in Fig. 4-3. The contribution of the sides 3 and 4 to the line integral is proportional to the length Δh. In the limit $\Delta h \to 0$ the sides 1 and 2 lie just on either side of the boundary surface and the contribution of sides 3 and 4 to the line integral vanishes. The surface integral in Eq. (4-41) may be written

$$\int_S \frac{\partial \mathbf{B}}{\partial t} \cdot \boldsymbol{\nu}\, \Delta b\, \Delta h$$

which also vanishes in the limit $\Delta h \to 0$. The length Δb is sufficiently small that **E** is constant along each of sides 1 and 2. Thus, Eq. (4-41) becomes

$$\oint_C \mathbf{E} \cdot d\mathbf{l} = \boldsymbol{\tau} \cdot (\mathbf{E}_2 - \mathbf{E}_1) \Delta b = 0 \tag{4-42}$$

Using Eq. (4-40) and the relation (1-30) that

$$\boldsymbol{\nu} \times \mathbf{n} \cdot \mathbf{E} = \boldsymbol{\nu} \cdot \mathbf{n} \times \mathbf{E} \tag{4-43}$$

we find that Eq. (4-42) yields

$$\boldsymbol{\nu} \cdot \mathbf{n} \times (\mathbf{E}_2 - \mathbf{E}_1) = 0 \tag{4-44}$$

Since the orientation of the rectangular surface element is arbitrary, Eq. (4-44) must be independent of the direction of the unit vector $\boldsymbol{\nu}$, so that

$$\mathbf{n} \times (\mathbf{E}_2 - \mathbf{E}_1) = 0 \tag{4-45}$$

Writing **E** in terms of its components along **n**, $\boldsymbol{\tau}$, and $\boldsymbol{\nu}$,

$$\mathbf{E} = \mathbf{n}E_n + \boldsymbol{\tau}E_\tau + \boldsymbol{\nu}E_\nu \tag{4-46}$$

we find from Eq. (4-45) that

$$\boldsymbol{\nu}(E_{\tau 2} - E_{\tau 1}) - \boldsymbol{\tau}(E_{\nu 2} - E_{\nu 1}) = 0 \tag{4-47}$$

or

$$\begin{aligned} E_{\tau 2} &= E_{\tau 1} \\ E_{\nu 2} &= E_{\nu 1} \end{aligned} \tag{4-48}$$

Thus, we conclude that the tangential components of **E** are continuous through any surface of discontinuity.

In a similar matter we may show from Eq. (4-11) that

$$\begin{aligned} &\text{(CGS)} \qquad \mathbf{n} \times (\mathbf{H}_2 - \mathbf{H}_1) = \frac{4\pi}{c}\mathbf{K} \\ &\text{(SI)} \qquad \mathbf{n} \times (\mathbf{H}_2 - \mathbf{H}_1) = \mathbf{K} \end{aligned} \tag{4-49}$$

where **K** is the surface current density on the boundary between the two media. If **K** is zero, then the tangential components of **H** are likewise continuous across the boundary.

4-7 Conservation of energy

In Secs. 4-3 and 4-4 we discussed the conservation of electric charge. The conservation of electromagnetic energy may also be derived from Maxwell's equations.

We dot the vector field **H** into Eq. (4-10) and the vector field $-\mathbf{E}$ into Eq. (4-11) and add the results to obtain

$$\text{(CGS)}\quad \mathbf{H}\cdot(\nabla\times\mathbf{E})-\mathbf{E}\cdot(\nabla\times\mathbf{H})+\frac{\mu}{c}\mathbf{H}\cdot\frac{\partial\mathbf{H}}{\partial t}+\frac{\epsilon}{c}\mathbf{E}\cdot\frac{\partial\mathbf{E}}{\partial t}=-\frac{4\pi}{c}\mathbf{E}\cdot\mathbf{J}$$

$$\text{(SI)}\quad \mathbf{H}\cdot(\nabla\times\mathbf{E})-\mathbf{E}\cdot(\nabla\times\mathbf{H})+\mu\mu_0\mathbf{H}\cdot\frac{\partial\mathbf{H}}{\partial t}+\epsilon\epsilon_0\mathbf{E}\cdot\frac{\partial\mathbf{E}}{\partial t}=-\mathbf{E}\cdot\mathbf{J}\qquad(4\text{-}50)$$

where the constitutive relations (4-25) and (4-26) have been used to eliminate **D** and **B**. From Eq. (3-46) we have

$$\nabla\cdot(\mathbf{E}\times\mathbf{H})=\mathbf{H}\cdot(\nabla\times\mathbf{E})-\mathbf{E}\cdot(\nabla\times\mathbf{H})\qquad(4\text{-}51)$$

Substituting this relation into Eqs. (4-50) and using Eq. (3-8), we obtain

$$\text{(CGS)}\quad \frac{1}{8\pi}\frac{\partial}{\partial t}(\epsilon E^2+\mu H^2)+\frac{c}{4\pi}\nabla\cdot(\mathbf{E}\times\mathbf{H})=-\mathbf{E}\cdot\mathbf{J}$$

$$\text{(SI)}\quad \tfrac{1}{2}\frac{\partial}{\partial t}(\epsilon\epsilon_0E^2+\mu\mu_0H^2)+\nabla\cdot(\mathbf{E}\times\mathbf{H})=-\mathbf{E}\cdot\mathbf{J}\qquad(4\text{-}52)$$

Equations (4-52) may be readily cast into the form of a conservation equation by letting

$$\text{(CGS)}\quad W=\frac{1}{8\pi}(\epsilon E^2+\mu H^2)$$

$$\text{(SI)}\quad W=\tfrac{1}{2}(\epsilon\epsilon_0E^2+\mu\mu_0H^2)\qquad(4\text{-}53)$$

be the volume density of electromagnetic energy and letting

$$\text{(CGS)}\quad \mathbf{S}=\frac{c}{4\pi}(\mathbf{E}\times\mathbf{H})$$

$$\text{(SI)}\quad \mathbf{S}=\mathbf{E}\times\mathbf{H}\qquad(4\text{-}54)$$

be the current density of energy flow. The field **S** is known as the *Poynting vector.* Then Eq. (4-52) becomes [cf. Eq. (4-6)]

$$\frac{\partial W}{\partial t}=-\nabla\cdot\mathbf{S}-\mathbf{E}\cdot\mathbf{J}\qquad(4\text{-}55)$$

which states that the rate of change of energy per unit volume equals the rate of flow of energy through the surface plus the internal production of energy per unit volume. Since the outward direction from a volume element is regarded as positive, **S** represents an energy flow out of the volume element and $-\mathbf{S}$ results in a positive $\partial W/\partial t$. The product $\mathbf{E}\cdot\mathbf{J}$ (which equals σE^2) is just the Joule heat per unit volume leaking in unit time out of the electromagnetic field. Thus, $-\mathbf{E}\cdot\mathbf{J}$ is a gain to the volume element and increases $\partial W/\partial t$. Hence, the two minus signs on the right-hand side of Eq. (4-55) are explained.

4-8 The wave equation

We consider a region of space with no charges or electric currents, so that $\rho = 0$ and $\mathbf{J} = 0$. In such an isotropic, homogeneous medium characterized by ϵ and μ, Maxwell's equations become

(CGS)	(SI)	
$\nabla \times \mathbf{E} + \dfrac{\mu}{c}\dfrac{\partial \mathbf{H}}{\partial t} = 0$	$\nabla \times \mathbf{E} + \mu\mu_0 \dfrac{\partial \mathbf{H}}{\partial t} = 0$	(4-56)
$\nabla \times \mathbf{H} - \dfrac{\epsilon}{c}\dfrac{\partial \mathbf{E}}{\partial t} = 0$	$\nabla \times \mathbf{H} - \epsilon\epsilon_0 \dfrac{\partial \mathbf{E}}{\partial t} = 0$	(4-57)
$\nabla \cdot \mathbf{E} = 0$	$\nabla \cdot \mathbf{E} = 0$	(4-58)
$\nabla \cdot \mathbf{H} = 0$	$\nabla \cdot \mathbf{H} = 0$	(4-59)

If we take the curl of Eq. (4-56) and then eliminate $\nabla \times \mathbf{H}$ by means of Eq. (4-57), we obtain

$$\text{(CGS)} \quad \nabla \times (\nabla \times \mathbf{E}) + \frac{\mu}{c}\frac{\partial}{\partial t}(\nabla \times \mathbf{H}) = \nabla \times (\nabla \times \mathbf{E}) + \frac{\epsilon\mu}{c^2}\frac{\partial^2 \mathbf{E}}{\partial t^2} = 0$$

$$\text{(SI)} \quad \nabla \times (\nabla \times \mathbf{E}) + \mu\mu_0\frac{\partial}{\partial t}(\nabla \times \mathbf{H}) = \nabla \times (\nabla \times \mathbf{E}) + \epsilon\epsilon_0\mu\mu_0\frac{\partial^2 \mathbf{E}}{\partial t^2} = 0 \tag{4-60}$$

Noting that from Eq. (3-45)

$$\nabla \times (\nabla \times \mathbf{E}) = \nabla(\nabla \cdot \mathbf{E}) - \nabla^2\mathbf{E} \tag{4-61}$$

and that, according to Eq. (4-58), $\nabla \cdot \mathbf{E}$ vanishes, we may write Eqs. (4-60) in the form

$$\text{(CGS)} \quad \nabla^2\mathbf{E} - \frac{\epsilon\mu}{c^2}\frac{\partial^2 \mathbf{E}}{\partial t^2} = 0$$

$$\text{(SI)} \quad \nabla^2\mathbf{E} - \epsilon\epsilon_0\mu\mu_0\frac{\partial^2 \mathbf{E}}{\partial t^2} = 0 \tag{4-62}$$

By taking the curl of Eq. (4-57) and using Eqs. (4-56) and (4-59), we obtain in a similar way

$$\text{(CGS)} \quad \nabla^2\mathbf{H} - \frac{\epsilon\mu}{c^2}\frac{\partial^2 \mathbf{H}}{\partial t^2} = 0$$

$$\text{(SI)} \quad \nabla^2\mathbf{H} - \epsilon\epsilon_0\mu\mu_0\frac{\partial^2 \mathbf{H}}{\partial t^2} = 0 \tag{4-63}$$

Thus, the components of **E** and of **H** obey the *wave equation,*

$$\nabla^2 u - \frac{1}{v^2}\frac{\partial^2 u}{\partial t^2} = 0 \qquad (4\text{-}64)$$

where u stands for E_x, E_y, E_z, H_x, H_y, or H_z and v is the velocity of wave u:

$$\begin{aligned} &\text{(CGS)} \qquad v = c/(\epsilon\mu)^{1/2} \\ &\text{(SI)} \qquad v = (\epsilon\epsilon_0\mu\mu_0)^{-1/2} \end{aligned} \qquad (4\text{-}65)$$

In a vacuum, ϵ and μ are each unity and v equals c, the velocity of light in free space ($c = 2.998 \times 10^{10}$ cm/s $= 2.998 \times 10^8$ m/s). Thus we obtain the important result that $\epsilon_0\mu_0 = c^{-2}$ in SI units. Since ϵ_0 has the value $(10^7/4\pi c^2)$ F/m (kg^{-1} m^{-3} s^4 A^2) in SI units as the result of the definition of the ampere (see Sec. 4-2), the value of μ_0 in SI units must be $4\pi \times 10^{-7}$ H/m (kg m s^{-2} A^{-2}) as introduced in Sec. 4-5. In a transparent material, the velocity of an electromagnetic (light) wave is less than c and is given by c/n, where $n(n>1)$ is the *index of refraction.* By comparison with Eq. (4-65), we see that

$$n = (\epsilon\mu)^{1/2} \qquad (4\text{-}66)$$

4-9 Plane electromagnetic waves

Generally, a wave u which is a solution of Eq. (4-64) is a function of the three space variables and the time: $u = u(x,y,z,t)$. We consider here a solution of the wave equation (4-64) where u depends on only one space variable, say z, and the time $t : u = u(z,t)$. Under this special condition, Eq. (4-64) reduces to

$$\frac{\partial^2 u}{\partial z^2} - \frac{1}{v^2}\frac{\partial^2 u}{\partial t^2} = 0 \qquad (4\text{-}67)$$

Any function $u(\zeta)$ of the variable $\zeta = z \pm vt + \alpha$ (where α is a constant called the *phase*) is a solution of the partial differential equation (4-67). If we follow the motion of a point determined by $\zeta = \alpha$, so that $z = \pm vt$, then the distance z traveled by that point is equal to the product of the velocity $\pm v$ and the time t. Thus, the solution with $\zeta = z - vt + \alpha$ represents a wave traveling in the positive z direction, and $\zeta = z + vt + \alpha$ gives a wave propagating in the negative z direction.

We consider a wave traveling in the positive z direction with $\alpha = 0$ and select the solution in terms of the cosine function; hence,

$$\begin{aligned} u(z,t) &= u_0 \cos k(z - vt) \\ &= u_0 \cos (kz - \omega t) \\ &= u_0 \cos 2\pi\left(\frac{z}{\lambda} - \nu t\right) \end{aligned} \qquad (4\text{-}68)$$

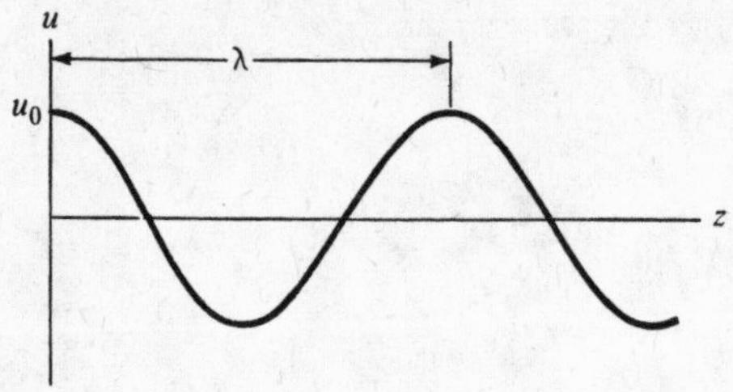

FIGURE 4-4

where u_0 is the value of $u(z, t)$ at $z = 0, t = 0$, and k is called the wave number. The angular frequency ω is equal to $2\pi\nu$, where ν is the number of oscillations per second as the light wave (electromagnetic wave) passes a fixed point. The wavelength λ is the distance between two neighboring crests of the wave. These various quantities are related by

$$k = \omega/v = 2\pi\nu/v$$
$$\lambda = v/\nu = 2\pi/k \tag{4-69}$$

A *plane wave* as given by Eq. (4-68) is illustrated in Fig. 4-4.

We recall that $u(z,t)$ in Eq. (4-68) refers to the components of **E** and of **H**. Therefore, the Maxwell equation (4-58) for a plane wave (4-68) becomes

$$\frac{\partial E_x}{\partial x} + \frac{\partial E_y}{\partial y} + \frac{\partial E_z}{\partial z} = 0 + 0 - E_{z0}\, k \sin (kz - \omega t) = 0$$

Since neither k nor $\sin (kz - \omega t)$ is identically zero, the constant E_{z0} must vanish and the z component E_z of the electric field vector **E** is always zero. Similarly, from Eq. (4-59) we find that H_z also vanishes, so that neither the electric nor the magnetic field has a component in the direction of propagation of the plane wave. The vectors **E** and **H** are therefore said to be *transverse.*

The Maxwell equations (4-56) and (4-57) provide further restrictions on **E** and **H**. The x component of Eq. (4-56) is

(CGS) $$\frac{\partial E_z}{\partial y} - \frac{\partial E_y}{\partial z} = -\frac{\mu}{c}\frac{\partial H_x}{\partial t}$$

(SI) $$\frac{\partial E_z}{\partial y} - \frac{\partial E_y}{\partial z} = -\mu\mu_0\frac{\partial H_x}{\partial t}$$

Substitution of the plane wave (4-68) yields

(CGS) $$0 + E_{y0}\, k \sin (kz - \omega t) = \frac{\mu}{c} H_{x0}\, (-\omega) \sin (kz - \omega t)$$

(SI) $$0 + E_{y0}\, k \sin (kz - \omega t) = \mu\mu_0 H_{x0}\, (-\omega) \sin (kz - \omega t)$$

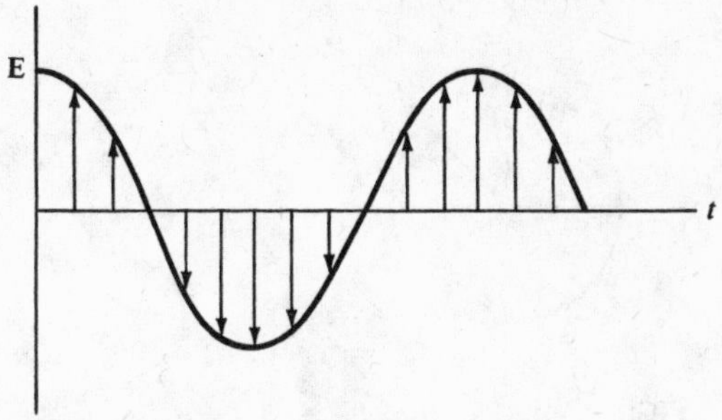

FIGURE 4-5

or

$$H_{x0} = -(\epsilon\epsilon_0/\mu\mu_0)^{1/2}E_{y0} \tag{4-70}$$

where Eqs. (4-65) and (4-69) have been introduced. Equation (4-70) is valid in both sets of units, since $\epsilon_0 = \mu_0 = 1$ in CGS units. From the y component of Eq. (4-56) we find that

$$H_{y0} = (\epsilon\epsilon_0/\mu\mu_0)^{1/2}E_{x0} \tag{4-71}$$

The same information may be obtained from the x and y components of Eq. (4-57). Thus, **E** and **H** for the plane wave are

$$\begin{aligned} \mathbf{E} &= (\mathbf{i}E_{x0} + \mathbf{j}E_{y0}) \cos(kz - \omega t) \\ \mathbf{H} &= (-\mathbf{i}E_{y0} + \mathbf{j}E_{x0})(\epsilon\epsilon_0/\mu\mu_0)^{1/2} \cos(kz - \omega t) \end{aligned} \tag{4-72}$$

One can readily see that $\mathbf{E} \cdot \mathbf{H} = 0$, so that **E** is perpendicular to **H**. Both are perpendicular to **k**, the direction of propagation.

4-10 Linear and circular polarization

In the preceding section we considered a plane-wave solution to the partial differential equation (4-67) with the phase α for each component of **E** and of **H** set equal to zero. We consider now a more general situation in which the phases for E_x and E_y are not the same. Again choosing the solution in terms of the cosine function, we may write the electric field for the transverse plane wave as

$$\mathbf{E} = \mathbf{i}E_{x0} \cos(kz - \omega t + \alpha_x) + \mathbf{j}E_{y0} \cos(kz - \omega t + \alpha_y) \tag{4-73}$$

The magnetic field **H** is perpendicular to **E** and to the unit vector **k** and its components are given by Eqs. (4-70) and (4-71). Therefore, a consideration of **E** automatically specifies **H**.

If α_x equals α_y, the ratio of the x and y components of **E** is

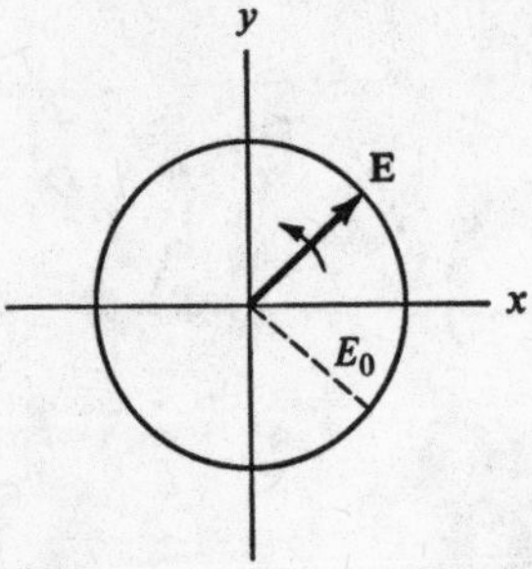

FIGURE 4-6

$$\frac{E_x}{E_y} = \frac{E_{x0}}{E_{y0}} = \text{const}$$

Since this ratio remains constant, the vector field **E** points in the same direction at each point in space and this direction does not change with time (see Fig. 4-5). Such a plane wave is said to be *linearly polarized.*

We next consider the situation where α_x and α_y differ by $\pm\pi/2$ ($\pm 90°$). We let α_x equal δ, so that α_y equals $\delta \pm \pi/2$. Moreover, we shall let E_{x0} and E_{y0} be the same:

$$E_{x0} = E_{y0} \equiv E_0$$

Since

$$\cos(\theta \pm \phi) = \cos\theta \cos\phi \mp \sin\theta \sin\phi \tag{4-74}$$

and $\sin \pi/2 = 1$, $\cos \pi/2 = 0$, we have

$$\cos(kz - \omega t + \delta \pm \pi/2) = \mp \sin(kz - \omega t + \delta)$$

so that

$$\mathbf{E} = \mathbf{i}E_0 \cos(kz - \omega t + \delta) \mp \mathbf{j}E_0 \sin(kz - \omega t + \delta) \tag{4-75}$$

The magnitude of the vector **E** in Eq. (4-75) (obtained from **E** · **E**) is a constant E_0. However, the ratio E_x/E_y is not a constant but equals

$$\frac{E_x}{E_y} = \frac{\cos(kz - \omega t + \delta)}{\mp \sin(kz - \omega t + \delta)} = \mp \cot(kz - \omega t + \delta)$$

Thus, the direction of **E** varies from point to point and changes with time.

If we square the x and y components of **E** and add, we obtain

$$E_x^2 + E_y^2 = E_0^2 \tag{4-76}$$

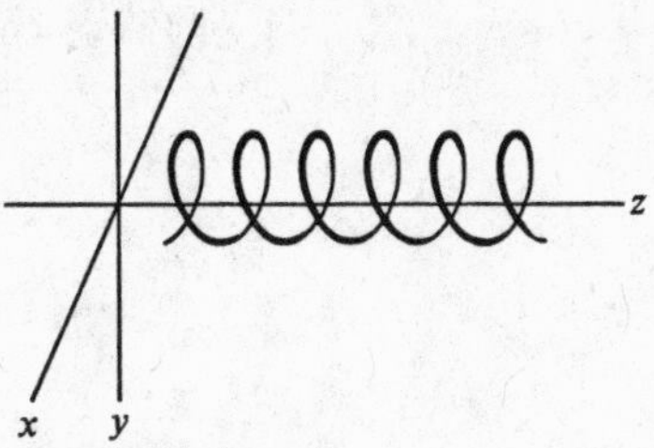

FIGURE 4-7*a*

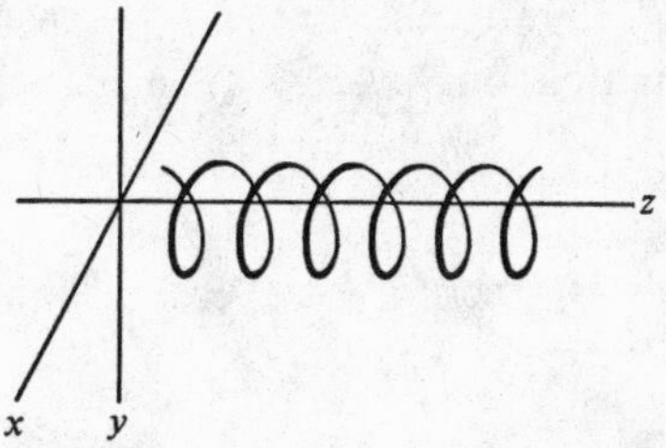

FIGURE 4-7*b*

which is the equation for a circle. For z remaining constant, the tip of the vector $\mathbf{E}$ sweeps a circle of radius E_0 at a frequency $\pm\omega$ as shown in Fig. 4-6. Since z is also varying, the vector $\mathbf{E}$ actually traces a cylindrical helix as illustrated in Fig. 4-7.

A plane wave as given by Eq. (4-75) is called *circularly polarized.* For the upper sign (−) the rotation of $\mathbf{E}$ is counterclockwise when the observer is looking at the oncoming wave (looking in the negative z direction). The vector $\mathbf{E}$ then traces a right-handed helix as in Fig. 4-7*a*. In optics such a wave is said to be *left circularly polarized.* The lower sign (+) gives a *right circularly polarized* wave (see Fig. 4-7*b*).

Linearly polarized light may be regarded as the superposition of two circularly polarized plane waves, a right circularly polarized wave with phase $+\delta$ and a left circularly polarized wave with phase $-\delta$. This equivalence may be readily shown by adding the right circularly polarized wave

$$\mathbf{E}_{\text{right}} = \mathbf{i}E_0 \cos(kz - \omega t + \delta) + \mathbf{j}E_0 \sin(kz - \omega t + \delta)$$

to the left circularly polarized wave

$$\mathbf{E}_{\text{left}} = \mathbf{i}E_0 \cos(kz - \omega t - \delta) - \mathbf{j}E_0 \sin(kz - \omega t - \delta)$$

The result is

$$\mathbf{E} = \mathbf{E}_{\text{right}} + \mathbf{E}_{\text{left}} = \mathbf{i}2E_0 \cos(kz - \omega t)\cos\delta + \mathbf{j}2E_0 \cos(kz - \omega t)\sin\delta \qquad (4\text{-}77)$$

where Eq. (4-74) and the relation

$$\sin(\theta \pm \phi) = \sin\theta \cos\phi \pm \cos\theta \sin\phi \qquad (4\text{-}78)$$

have been used. Since the ratio E_x/E_y is a constant,

$$\frac{E_x}{E_y} = \frac{\cos\delta}{\sin\delta}$$

Eq. (4-77) is a linearly polarized plane wave.

Finally we consider the most general plane wave as given by Eq. (4-73) with no restrictions on $E_{x0}, E_{y0}, \alpha_x, \alpha_y$. Using Eq. (4-74), we write the components of **E** in Eq. (4-73) in the form

$$\frac{E_x}{E_{x0}} = \cos(kz - \omega t)\cos\alpha_x - \sin(kz - \omega t)\sin\alpha_x$$

$$\frac{E_y}{E_{y0}} = \cos(kz - \omega t)\cos\alpha_y - \sin(kz - \omega t)\sin\alpha_y \tag{4-79}$$

We eliminate first $\sin(kz - \omega t)$ and then $\cos(kz - \omega t)$ to obtain

$$\begin{aligned}\frac{E_x}{E_{x0}}\sin\alpha_y - \frac{E_y}{E_{y0}}\sin\alpha_x &= \cos(kz - \omega t)(\sin\alpha_y\cos\alpha_x - \cos\alpha_y\sin\alpha_x)\\ &= \cos(kz - \omega t)\sin\Delta\alpha\\ \frac{E_x}{E_{x0}}\cos\alpha_y - \frac{E_y}{E_{y0}}\cos\alpha_x &= \sin(kz - \omega t)(\sin\alpha_y\cos\alpha_x - \cos\alpha_y\sin\alpha_x)\\ &= \sin(kz - \omega t)\sin\Delta\alpha\end{aligned} \tag{4-80}$$

where

$$\Delta\alpha = \alpha_y - \alpha_x$$

Squaring Eqs. (4-80) and adding, we obtain

$$\left(\frac{E_x}{E_{x0}}\right)^2 + \left(\frac{E_y}{E_{y0}}\right)^2 - 2\left(\frac{E_x}{E_{x0}}\right)\left(\frac{E_y}{E_{y0}}\right)\cos\Delta\alpha = \sin^2\Delta\alpha \tag{4-81}$$

Equation (4-81) is the equation for an ellipse whose semimajor axis forms an angle χ with the x axis, such that

$$\tan 2\chi = \frac{2E_{x0}E_{y0}}{E_{x0}^2 - E_{y0}^2}\cos\Delta\alpha$$

Thus, the electric field vector **E** of the plane wave traces an ellipse as shown in Fig. 4-8. Such a plane wave is said to be *elliptically polarized.*

When $\Delta\alpha = \pi/2$ and $E_{x0} = E_{y0}$, Eq. (4-81) becomes identical with Eq. (4-76) and the ellipse becomes a circle. When $\Delta\alpha = 0$ or π (180°), Eq. (4-81) gives

$$\frac{E_x}{E_y} = \pm\frac{E_{x0}}{E_{y0}}$$

so that the ellipse degenerates into a straight line and the wave is linearly polarized.

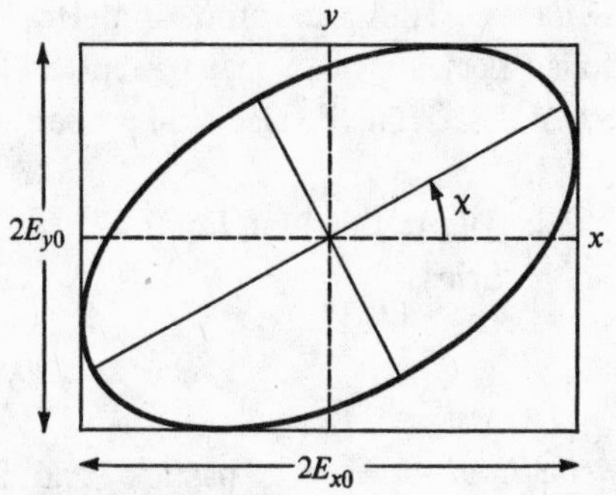

FIGURE 4-8

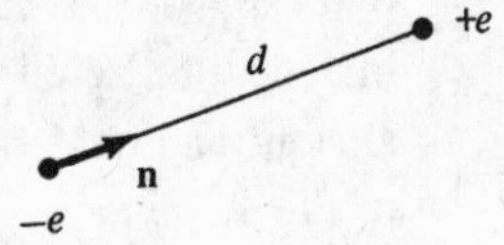

FIGURE 4-9

4-11 Electric and magnetic moments

Until now we have considered only matter in bulk and have not introduced the molecular parameters which influence the electrical and magnetic properties of a medium. Actually, bulk matter is composed of atoms or molecules, which consist of one or more positively charged nuclei and of negatively charged electrons. When such an atom or molecule is placed in an external electric field **E**, each charged particle (nuclei and electrons) experiences a force according to Eq. (4-9). The forces on the positive and negative charges are oppositely directed and therefore displace the charges in opposite directions. These charges are displaced to the extent that the forces due to the applied field **E** are balanced by the internal forces holding the atom or molecule together.

The existence of this charge displacement is the reason why **D** differs from **E**. This difference is related, according to Eqs. (4-29), to the electric polarization **P**. If there are N atoms or molecules in a volume V, we may regard **P** as the sum of contributions from each of the N particles,

$$\mathbf{P} = \frac{1}{V} \sum_{i=1}^{N} \mathbf{p}_i \tag{4-82}$$

where $\mathbf{p}_i$ is the atomic or molecular electric moment of particle i. If we let **p** be the average particle electric moment, then **P** is

$$\mathbf{P} = \frac{N}{V} \mathbf{p} \tag{4-83}$$

If two charges, one $+e$ and the other $-e$ in magnitude, are separated by a distance d, they make a contribution

$$\mathbf{p} = ed\mathbf{n} \tag{4-84}$$

to the polarization. The unit vector **n** is directed from the negative charge to the positive charge (see Fig. 4-9). Such a configuration is called an electric dipole, and

p in Eq. (4-84) is called the electric dipole moment. Thus, each atom or molecule in a medium acted upon by an external applied electric field **E** may be replaced by an equivalent electric dipole for the purpose of describing the electrical properties of the medium.

In a similar way we may regard the magnetic polarization **M** in Eqs. (4-29) as the sum of contributions from each of the N particles,

$$\mathbf{M} = \frac{1}{V} \sum_{i=1}^{N} \mathbf{m}_i = \frac{N}{V}\,\mathbf{m} \tag{4-85}$$

where $\mathbf{m}_i$ is the *magnetic moment* or *magnetic dipole moment* of particle i and **m** is the average particle magnetic moment. The magnetic dipole moment of an atom or molecule arises from the motion of charges within the particle. These movements of charge or currents $\mathbf{j}(\mathbf{r})$ give a magnetic moment according to the relation

$$\begin{aligned} &(\text{CGS}) \qquad \mathbf{m} = \frac{1}{2c} \int_{V_0} \mathbf{r} \times \mathbf{j}(\mathbf{r})\, dv \\ &(\text{SI}) \qquad \mathbf{m} = \tfrac{1}{2} \int_{V_0} \mathbf{r} \times \mathbf{j}(\mathbf{r})\, dv \end{aligned} \tag{4-86}$$

where V_0 is the volume of the atom or molecule.

The electric field acting on a particle in a dense medium (such as a liquid or a solid) is not equal to just the externally applied electric field **E**. The electric moments of the neighboring particles also contribute to the effective electric field on this specific particle. Thus, within the medium the effective field $\mathbf{E}'$ is the sum of the external field **E** and a polarization field which is proportional to **P**. For a randomly oriented distribution of electric dipoles **p** within the medium, the polarization field may be shown to be $4\pi\mathbf{P}/3$ in CGS units or $\mathbf{P}/3\epsilon_0$ in SI units, so that

$$\begin{aligned} &(\text{CGS}) \qquad \mathbf{E}' = \mathbf{E} + (4\pi\mathbf{P}/3) \\ &(\text{SI}) \qquad \mathbf{E}' = \mathbf{E} + (\mathbf{P}/3\epsilon_0) \end{aligned} \tag{4-87}$$

It is observed that the average electric dipole moment **p** induced in a particle by an external electric field is proportional to the effective electric field acting on the particle:

$$\mathbf{p} = \alpha \mathbf{E}' \tag{4-88}$$

The proportionality constant α is known as the atomic or molecular *polarizability*. Substitution of Eq. (4-88) into Eq. (4-83) and application of Eqs. (4-87) yield

(CGS) $\quad \mathbf{P} = \frac{N}{V} \alpha \left(\mathbf{E} + \frac{4\pi}{3}\mathbf{P}\right)$

(SI) $\quad \mathbf{P} = \frac{N}{V} \alpha \left(\mathbf{E} + \frac{\mathbf{P}}{3\epsilon_0}\right)$

From Eqs. (4-30) we see that

$$\text{(CGS)} \quad \frac{\epsilon - 1}{\epsilon + 2} = \frac{4\pi N\alpha}{3V} \qquad \text{(SI)} \quad \frac{\epsilon - 1}{\epsilon + 2} = \frac{N\alpha}{3\epsilon_0 V} \tag{4-89}$$

and

$$\text{(CGS)} \quad \chi_e = \frac{N\alpha/V}{1 - (4\pi N\alpha/3V)} \qquad \text{(SI)} \quad \chi_e = \frac{N\alpha/V}{\epsilon_0 - (N\alpha/3V)} \tag{4-90}$$

Thus, the dielectric constant and electric susceptibility are related to the polarizability. Equations (4-89) are known as the Clausius-Mossotti relation.

Although similar expressions could be developed for the magnetic field, for most substances the effective magnetic field acting on an atom or molecule is just the externally applied field. Therefore, such a development is unnecessary.

4-12 Optical activity

The ability of a medium to rotate the plane of polarization of linearly polarized light which is transmitted through it is called *optical rotatory power.* The observed optical rotatory power of a substance is usually expressed in terms of the specific rotation $[\alpha]$ or the molecular rotation [M]. For an optically active substance in solution, the specific rotation $[\alpha]$ is dependent on the frequency ν of the incident linearly polarized light and is defined as the observed angle of rotation in degrees for a 1 dm path length and for a concentration of 1 g/cm^3. When expressed in radians and either CGS or SI units, this definition of $[\alpha]$ becomes

$$\text{(CGS)} \quad [\alpha]_\nu = \frac{1800\, V\phi}{\pi M} \qquad \text{(SI)} \quad [\alpha]_\nu = \frac{18{,}000\, V\phi}{\pi M} \tag{4-91}$$

where M is the molar mass (molecular weight in CGS units and kilograms per mole in SI units) and V is the volume occupied in solution by a mole of the optically active substance. The angle ϕ is the observed angle of rotation of the linearly polarized light of frequency ν and is expressed in radians per centimeter in CGS units and radians per meter in SI units. The molecular rotation [M] is

$$\begin{aligned} &\text{(CGS)} \qquad [\mathrm{M}]_\nu = \frac{M[\alpha]_\nu}{100} = \frac{18V\phi}{\pi} \\ &\text{(SI)} \qquad [\mathrm{M}]_\nu = 10M[\alpha]_\nu = \frac{180{,}000V\phi}{\pi} \end{aligned} \tag{4-92}$$

We now apply the electromagnetic theory of light to optical activity. An isotropic solution of asymmetric molecules is optically active if the induced electric moment **p** and the induced magnetic moment **m** of a single molecule can be written in the form

$$\begin{aligned} \mathbf{p} &= \alpha\mathbf{E}' - \beta\mu_0\frac{\partial\mathbf{H}}{\partial t} \\ \mathbf{m} &= \beta\frac{\partial\mathbf{E}'}{\partial t} \end{aligned} \tag{4-93}$$

where α and $\mathbf{E}'$ are discussed in Sec. 4-11, $\mu_0 = 1$ in CGS units, and β is called the molecule rotatory parameter. The assumptions that the effective magnetic field is just **H** and that the relative permeability μ is unity have been made here. Expressions of the form of Eqs. (4-93) have been derived from quantum theory, and an explicit formula for β in terms of electronic wave functions has been obtained.[1] Substitution of Eqs. (4-83), (4-85), (4-87), (4-89), and (4-93) into Eqs. (4-29) gives

$$\mathbf{D} = \epsilon\epsilon_0\mathbf{E} - \gamma\mu_0\frac{\partial\mathbf{H}}{\partial t} \tag{4-94}$$

$$\mathbf{B} = \mu_0\left(\mathbf{H} + \gamma\frac{\partial\mathbf{E}}{\partial t}\right) \tag{4-95}$$

where, as before, $\epsilon_0 = \mu_0 = 1$ in CGS units and where

$$\begin{aligned} &\text{(CGS)} \qquad \gamma = \frac{4\pi N}{V}\,\frac{\epsilon+2}{3}\,\beta \\ &\text{(SI)} \qquad \gamma = \frac{N}{V}\,\frac{\epsilon+2}{3}\,\beta \end{aligned} \tag{4-96}$$

The quantity N now refers to the number of molecules per mole (Avogadro's number). Thus, the simple constitutive relations (4-25) and (4-26) with $\mu = 1$ do not hold for this situation. If, however, β vanishes, then Eqs. (4-94) and (4-95) reduce to Eqs. (4-25) and (4-26).

[1] See, for example, E. U. Condon, *Rev. Mod. Phys.*, **9**:432 (1937).

The Maxwell relations (4-10) to (4-13) in the absence of electric charges and currents become

(CGS)	(SI)	
$\nabla \times \mathbf{E} = -\frac{1}{c}\frac{\partial \mathbf{B}}{\partial t}$	$\nabla \times \mathbf{E} = -\frac{\partial \mathbf{B}}{\partial t}$	(4-97*a*)
$\nabla \times \mathbf{H} = \frac{1}{c}\frac{\partial \mathbf{D}}{\partial t}$	$\nabla \times \mathbf{H} = \frac{\partial \mathbf{D}}{\partial t}$	(4-97*b*)
$\nabla \cdot \mathbf{D} = 0$	$\nabla \cdot \mathbf{D} = 0$	(4-97*c*)
$\nabla \cdot \mathbf{B} = 0$	$\nabla \cdot \mathbf{B} = 0$	(4-97*d*)

By the same arguments presented in Secs. 4-8 and 4-9, the plane-wave solution of Eqs. (4-97) gives **D** and **B** as transverse waves ($D_z = 0$, $B_z = 0$). The z components of Eqs. (4-94) and (4-95) when solved simultaneously yield the fact that $E_z = 0$ and $H_z = 0$.

We now consider a right circularly polarized plane wave, for which **E** is given by Eq. (4-75), and initially set δ equal to zero:

$$\mathbf{E} = E_0 (\mathbf{i} \cos \psi + \mathbf{j} \sin \psi) \tag{4-98}$$

where

$$\psi = kz - \omega t = \omega\left(\frac{nz}{c} - t\right)$$

with n the index of refraction. We wish to find the corresponding expressions for **H, D,** and **B**. We first take the curl of **E** in Eq. (4-98):

$$\nabla \times \mathbf{E} = -\mathbf{i}\frac{\partial E_y}{\partial z} + \mathbf{j}\frac{\partial E_x}{\partial z} = -\mathbf{i}E_0 \frac{\omega n}{c} \cos \psi - \mathbf{j}E_0 \frac{\omega n}{c} \sin \psi$$

From Eq. (4-97*a*) it follows by integration of $\nabla \times \mathbf{E}$ with respect to time that

$$\text{(CGS)} \qquad \mathbf{B} = nE_0 (-\mathbf{i} \sin \psi + \mathbf{j} \cos \psi)$$
$$\text{(SI)} \qquad \mathbf{B} = \frac{nE_0}{c}(-\mathbf{i} \sin \psi + \mathbf{j} \cos \psi) \tag{4-99}$$

Since Eqs. (4-98) and (4-99) give

$$\text{(CGS)} \qquad \frac{\partial \mathbf{E}}{\partial t} = \omega E_0(\mathbf{i} \sin \psi - \mathbf{j} \cos \psi) = -\frac{\omega \mathbf{B}}{n}$$

$$\text{(SI)} \qquad \frac{\partial \mathbf{E}}{\partial t} = \omega E_0(\mathbf{i} \sin \psi - \mathbf{j} \cos \psi) = -\frac{\omega c \mathbf{B}}{n}$$

Eq. (4-95) becomes

$$\text{(CGS)} \qquad \mathbf{H} = \left(1 + \frac{\omega\gamma}{n}\right)\mathbf{B} = (n + \omega\gamma)E_0(-\mathbf{i}\sin\psi + \mathbf{j}\cos\psi)$$

(4-100)

$$\text{(SI)} \qquad \mathbf{H} = \left(\frac{1}{\mu_0} + \frac{\omega c\gamma}{n}\right)\mathbf{B} = (n + \mu_0 c\omega\gamma)\frac{E_0}{\mu_0 c}(-\mathbf{i}\sin\psi + \mathbf{j}\cos\psi)$$

Integrating the curl of **H** in Eqs. (4-100) with respect to time and using Eq. (4-97*b*), we obtain

$$\text{(CGS)} \qquad \mathbf{D} = (n + \omega\gamma)\, nE_0(\mathbf{i}\cos\gamma + \mathbf{j}\sin\psi) = (n + \omega\gamma)\, n\mathbf{E}$$

(4-101)

$$\text{(SI)} \qquad \mathbf{D} = (n + \mu_0 c\omega\gamma)\frac{nE_0}{\mu_0 c^2}(\mathbf{i}\cos\psi + \mathbf{j}\sin\psi) = (n + \mu_0 c\omega\gamma)\frac{n}{\mu_0 c^2}\mathbf{E}$$

Thus we observe that **E** and **D** are parallel, **H** and **B** are parallel, and **E** and **H** are perpendicular.

If we substitute Eqs. (4-100) for **H** into Eq. (4-94) and use Eqs. (4-101), we obtain

$$\text{(CGS)} \qquad \mathbf{D} = \frac{\epsilon\mathbf{E}}{1 + (\omega\gamma/n)}$$

(4-102)

$$\text{(SI)} \qquad \mathbf{D} = \frac{\epsilon\epsilon_0\mathbf{E}}{1 + \mu_0 c\omega\gamma/n}$$

In order that Eqs. (4-101) and (4-102) be identical, we must have

$$\text{(CGS)} \qquad (n + \omega\gamma)^2 = \epsilon$$

(4-103)

$$\text{(SI)} \qquad (n + \mu_0 c\omega\gamma)^2 = \epsilon$$

where the relation $\epsilon_0\mu_0 = c^{-2}$ in SI units has been used. When solved for n, Eqs. (4-103) give

$$\text{(CGS)} \qquad n_r = \epsilon^{1/2} - \omega\gamma$$

(4-104)

$$\text{(SI)} \qquad n_r = \epsilon^{1/2} - \mu_0 c\omega\gamma$$

We have introduced the notation n_r for the index of refraction for right circularly polarized light.

For left circularly polarized light, **E** is given by Eq. (4-75) as

$$\mathbf{E} = E_0(\mathbf{i}\cos\psi - \mathbf{j}\sin\psi) \qquad (4\text{-}105)$$

Solving the Maxwell relations (4-97) for this case, we find that the index of refraction n_l for left circularly polarized light is

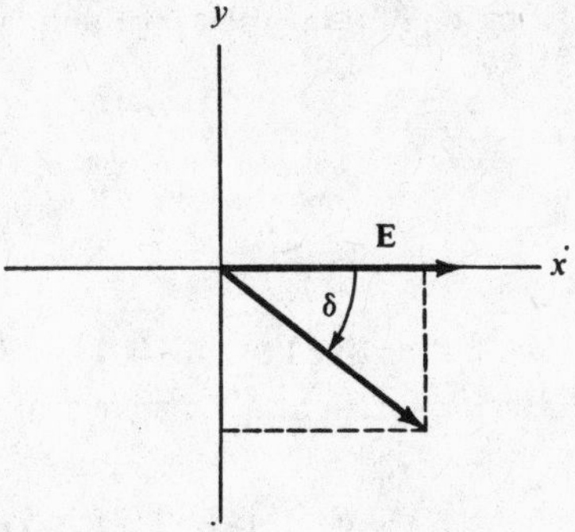

FIGURE 4-10

$$\text{(CGS)} \qquad n_l = \epsilon^{1/2} + \omega\gamma$$
$$\text{(SI)} \qquad n_l = \epsilon^{1/2} + \mu_0 c\omega\gamma \tag{4-106}$$

The angle ψ can, therefore, take two different forms, ψ_r and ψ_l,

$$\psi_r = \omega\left(\frac{n_r z}{c} - t\right) = \psi_0 - \delta$$
$$\psi_l = \omega\left(\frac{n_l z}{c} - t\right) = \psi_0 + \delta \tag{4-107}$$

where

$$\psi_0 = \frac{\omega\epsilon^{1/2} z}{c} - \omega t$$

and

$$\text{(CGS)} \qquad \delta = \omega^2\gamma z/c$$
$$\text{(SI)} \qquad \delta = \mu_0\omega^2\gamma z$$

Since a linearly polarized plane wave can be regarded as the superposition of right and left circularly polarized waves, we add Eqs. (4-98) and (4-105), using Eqs. (4-74), (4-78), and (4-107), to obtain

$$\mathbf{E} = \mathbf{i}E_0(\cos\psi_r + \cos\psi_l) + \mathbf{j}E_0(\sin\psi_r - \sin\psi_l)$$
$$= 2E_0 \cos\psi_0(\mathbf{i}\cos\delta - \mathbf{j}\sin\delta) \tag{4-108}$$

When $\delta = 0$, the y component of Eq. (4-108) vanishes and **E** is in the x direction. For $\delta > 0$, the vector **E** has rotated through the angle δ in a clockwise direction (see Fig. 4-10). [Recall that here δ is a function of z, whereas in Eq. (4-77) δ was a constant.] If light enters the medium at $z = 0$ and leaves at $z = d$, the angle of rotation of **E** in radians per unit length is, therefore,

$$\text{(CGS)} \qquad \phi = \frac{\delta}{d} = \frac{\omega^2\gamma}{c} = \frac{16\pi^3 Nc}{\lambda^2 V}\frac{\epsilon+2}{3}\beta$$

$$\text{(SI)} \qquad \phi = \frac{\delta}{d} = \mu_0\omega^2\gamma = \frac{4\pi^2 N}{\epsilon_0\lambda^2 V}\frac{\epsilon+2}{3}\beta \qquad \text{(4-109)}$$

Substituting Eqs. (4-109) into Eqs. (4-91) and setting $\epsilon = n^2$, we relate the specific rotation to the molecular rotatory parameter β,

$$\text{(CGS)} \qquad [\alpha]_\nu = 28{,}800\frac{\pi^2 Nc}{M\lambda^2}\frac{n_\nu^{\;2}+2}{3}\beta$$

$$\text{(SI)} \qquad [\alpha]_\nu = 72{,}000\frac{\pi N}{\epsilon_0 M\lambda^2}\frac{n_\nu^{\;2}+2}{3}\beta \qquad \text{(4-110)}$$

where the dependence of n upon the frequency ν has been noted.

4-13 Vector and scalar potentials

Maxwell's relations (4-10) to (4-13) are a set of coupled partial differential equations relating the components of **E, H, D,** and **B.** Rather than solve this set of equations for a particular problem, it is sometimes convenient to introduce potentials that satisfy the Maxwell relations and then to solve for these potentials.

Since **B** is solenoidal ($\nabla \cdot \mathbf{B} = 0$) and since the divergence of the curl of a vector function vanishes identically [Eq. (3-43)], we define a vector field **A** called the *vector potential* by the relation

$$\mathbf{B} = \nabla \times \mathbf{A} \qquad \text{(4-111)}$$

The Maxwell equation (4-10) becomes then

$$\text{(CGS)} \qquad \nabla \times \left(\mathbf{E} + \frac{1}{c}\frac{\partial \mathbf{A}}{\partial t}\right) = 0$$

$$\text{(SI)} \qquad \nabla \times \left(\mathbf{E} + \frac{\partial \mathbf{A}}{\partial t}\right) = 0$$

Since the curl of the gradient of a scalar function also vanishes identically [Eq. (3-44)], we may let

$$\text{(CGS)} \qquad \mathbf{E} + \frac{1}{c}\frac{\partial \mathbf{A}}{\partial t} = -\nabla\phi$$

$$\text{(SI)} \qquad \mathbf{E} + \frac{\partial \mathbf{A}}{\partial t} = -\nabla\phi \qquad \text{(4-112)}$$

where ϕ is the *scalar potential.*

The remaining two Maxwell relations determine the behavior of **A** and ϕ. Equation (4-11) becomes

$$\text{(CGS)} \qquad \nabla \times \mathbf{B} = \frac{\epsilon\mu}{c}\frac{\partial \mathbf{E}}{\partial t} + \frac{4\pi\mu}{c}\mathbf{J}$$

$$\text{(SI)} \qquad \nabla \times \mathbf{B} = \epsilon\epsilon_0\mu\mu_0 \frac{\partial \mathbf{E}}{\partial t} + \mu\mu_0\mathbf{J}$$

or

$$\text{(CGS)} \qquad \nabla \times (\nabla \times \mathbf{A}) = -\frac{\epsilon\mu}{c}\nabla\left(\frac{\partial\phi}{\partial t}\right) - \frac{\epsilon\mu}{c^2}\frac{\partial^2\mathbf{A}}{\partial t^2} + \frac{4\pi\mu}{c}\mathbf{J}$$

$$\text{(SI)} \qquad \nabla \times (\nabla \times \mathbf{A}) = -\epsilon\epsilon_0\mu\mu_0\nabla\left(\frac{\partial\phi}{\partial t}\right) - \epsilon\epsilon_0\mu\mu_0\frac{\partial^2\mathbf{A}}{\partial t^2} + \mu\mu_0\mathbf{J}$$

or

$$\text{(CGS)} \qquad \nabla^2\mathbf{A} - \frac{\epsilon\mu}{c^2}\frac{\partial^2\mathbf{A}}{\partial t^2} - \nabla\left[\nabla\cdot\mathbf{A} + \frac{\epsilon\mu}{c}\frac{\partial\phi}{\partial t}\right] = -\frac{4\pi\mu}{c}\mathbf{J}$$

$$\text{(SI)} \qquad \nabla^2\mathbf{A} - \epsilon\epsilon_0\mu\mu_0\frac{\partial^2\mathbf{A}}{\partial t^2} - \nabla\left[\nabla\cdot\mathbf{A} + \epsilon\epsilon_0\mu\mu_0\frac{\partial\phi}{\partial t}\right] = -\mu\mu_0\mathbf{J} \qquad \text{(4-113)}$$

where Eq. (3-45) has been introduced. Equation (4-12) takes the form

$$\text{(CGS)} \qquad \nabla^2\phi + \frac{1}{c}\frac{\partial\nabla\cdot\mathbf{A}}{\partial t} = \frac{-4\pi\rho}{\epsilon}$$

$$\text{(SI)} \qquad \nabla^2\phi + \frac{\partial\nabla\cdot\mathbf{A}}{\partial t} = -\frac{\rho}{\epsilon\epsilon_0}$$

or

$$\text{(CGS)} \qquad \nabla^2\phi - \frac{\epsilon\mu}{c^2}\frac{\partial^2\phi}{\partial t^2} + \frac{1}{c}\frac{\partial}{\partial t}\left[\nabla\cdot\mathbf{A} + \frac{\epsilon\mu}{c}\frac{\partial\phi}{\partial t}\right] = \frac{-4\pi\rho}{\epsilon}$$

$$\text{(SI)} \qquad \nabla^2\phi - \epsilon\epsilon_0\mu\mu_0\frac{\partial^2\phi}{\partial t^2} + \frac{\partial}{\partial t}\left[\nabla\cdot\mathbf{A} + \epsilon\epsilon_0\mu\mu_0\frac{\partial\phi}{\partial t}\right] = \frac{-\rho}{\epsilon\epsilon_0} \qquad \text{(4-114)}$$

In the definitions (4-111) and (4-112) of **A** and ϕ, the partial derivatives $(\partial A_x/\partial x)$, $(\partial A_y/\partial y)$, $(\partial A_z/\partial z)$, $(\partial\phi/\partial t)$ do not occur. Therefore, the scalar function ϕ and the components of **A** are not completely specified by these two equations. Consequently, we may further require that

$$\text{(CGS)} \qquad \frac{\partial A_x}{\partial x} + \frac{\partial A_y}{\partial y} + \frac{\partial A_z}{\partial z} + \frac{\epsilon\mu}{c}\frac{\partial\phi}{\partial t} = \nabla\cdot\mathbf{A} + \frac{\epsilon\mu}{c}\frac{\partial\phi}{\partial t} = 0$$

$$\text{(SI)} \qquad \frac{\partial A_x}{\partial x} + \frac{\partial A_y}{\partial y} + \frac{\partial A_z}{\partial z} + \epsilon\epsilon_0\mu\mu_0\frac{\partial\phi}{\partial t} = \nabla\cdot\mathbf{A} + \epsilon\epsilon_0\mu\mu_0\frac{\partial\phi}{\partial t} = 0 \qquad \text{(4-115)}$$

When this requirement is introduced, Eqs. (4-113) and (4-114) become

$$\text{(CGS)} \qquad \nabla^2\mathbf{A} - \frac{\epsilon\mu}{c^2}\frac{\partial^2\mathbf{A}}{\partial t^2} = \frac{-4\pi\mu}{c}\mathbf{J} \qquad \text{(SI)} \qquad \nabla^2\mathbf{A} - \epsilon\epsilon_0\mu\mu_0\frac{\partial^2\mathbf{A}}{\partial t^2} = -\mu\mu_0\mathbf{J} \tag{4-116}$$

$$\nabla^2\phi - \frac{\epsilon\mu}{c^2}\frac{\partial^2\phi}{\partial t^2} = \frac{-4\pi}{c}\rho \qquad \nabla^2\phi - \epsilon\epsilon_0\mu\mu_0\frac{\partial^2\phi}{\partial t^2} = \frac{-\rho}{\epsilon\epsilon_0} \tag{4-117}$$

We then have one partial differential equation for the vector potential **A** and another for the scalar potential ϕ. Equations (4-115) to (4-117) are equivalent to the Maxwell equations (4-10) to (4-13) and contain all the same relationships. Note that only partial derivatives of **A** and ϕ appear in these equations, so that to specify **A** and ϕ completely constants of integration must be determined by the boundary conditions for the specific problem being considered. If there are no charges or currents ($\rho = 0, \mathbf{J} = 0$), **A** and ϕ in Eqs. (4-116) and (4-117) obey the wave equation (4-64).

The relation (4-115) between **A** and ϕ is called the *Lorentz gauge.* Another useful gauge is the *Coulomb gauge* for which

$$\nabla \cdot \mathbf{A} = 0$$

Equations (4-113) and (4-114) for the Coulomb gauge are

$$\text{(CGS)} \qquad \nabla^2\mathbf{A} - \frac{\epsilon\mu}{c^2}\frac{\partial^2\mathbf{A}}{\partial t^2} = \frac{-4\pi\mu}{c}\mathbf{J} + \frac{\epsilon\mu}{c}\nabla\left(\frac{\partial\phi}{\partial t}\right)$$

$$\text{(SI)} \qquad \nabla^2\mathbf{A} - \epsilon\epsilon_0\mu\mu_0\frac{\partial^2\mathbf{A}}{\partial t^2} = -\mu\mu_0\mathbf{J} + \epsilon\epsilon_0\mu\mu_0\nabla\left(\frac{\partial\phi}{\partial t}\right) \tag{4-118}$$

and

$$\text{(CGS)} \qquad \nabla^2\phi = \frac{-4\pi\rho}{\epsilon}$$

$$\text{(SI)} \qquad \nabla^2\phi = \frac{-\rho}{\epsilon\epsilon_0} \tag{4-119}$$

Equation (4-119) is called Poisson's equation and reduces for a system with no charges to Laplace's equation

$$\nabla^2\phi = 0 \tag{4-120}$$

For some problems the Coulomb gauge is more convenient mathematically than the Lorentz gauge.

PROBLEMS

1. If a particle of mass *m* and charge *e* moves in a plane which is perpendicular to a constant magnetic field with induction **B**, show that the particle moves in a circle of radius mv/eB (SI units) with constant speed v.

2. Derive Eqs. (4-49) and show that the tangential components of **H** are continuous across a surface of discontinuity if the surface current density vanishes.

3. Show that the components of the field vectors **D** and **B** obey the wave equation (4-64).

4. Show that for a conductor with an electric current density **J** Eqs. (4-62) for **E** are replaced by

$$\text{(CGS)} \qquad \nabla^2 \mathbf{E} - \frac{4\pi\sigma\mu}{c^2}\frac{\partial \mathbf{E}}{\partial t} - \frac{\epsilon\mu}{c^2}\frac{\partial^2 \mathbf{E}}{\partial t^2} = 0$$

$$\text{(SI)} \qquad \nabla^2 \mathbf{E} - \mu\mu_0\sigma\frac{\partial \mathbf{E}}{\partial t} - \frac{\epsilon\mu}{c^2}\frac{\partial^2 \mathbf{E}}{\partial t^2} = 0$$

Find the corresponding expression for the magnetic field vector **H**.

5. If a particle with charge e moves with velocity v in the field of a plane electromagnetic wave, show that the ratio of the magnetic force to the coulombic force is of the order v/c.

6. Let E_0 be the amplitude of the electric field vector for a plane wave propagating in free space.

   (*a*) Show that the amplitudes of the magnetic vectors H_0 and B_0 are given in SI units by
   $H_0 = 2.654 \times 10^{-3}\, E_0$ ampere-turn/meter
   $B_0 = \frac{1}{3} \times 10^{-8}\, E_0$ weber/meter^2

   (*b*) When all quantities are expressed in CGS units, show that
   $H_0 = E_0$ oersted
   $B_0 = E_0$ gauss

7. Let E_0 be the amplitude of the electric field vector for a plane wave propagating in free space and $\bar{S}$ be the mean value of the energy flow vector **S** $[\bar{S} = (1/2\pi)\int_0^{2\pi} |\mathbf{S}|\, d\theta]$. Show that in SI units
   $\bar{S} = 1.327 \times 10^{-3}\, E_0^2$ watt/meter^2

8. The molecular rotation $[M]_D$ of many optically active molecules is of the order of 100° for the sodium D line ($\lambda = 0.589 \times 10^{-4}$ cm $= 589 \times 10^{-9}$ m $=$ 598 nm). A typical solution has an index of refraction of about 1.40.

   (*a*) Calculate the value of the molecular rotatory parameter β in both CGS and SI units.

(*b*) A typical value for the polarizability α is of the order of 10^{-22} cm = 10^{-28} m^3. Compare the size of the dipole moment **p** induced in the optically active molecule by the changing magnetic field with that induced by the electric field **E** by finding the ratio of these two contributions.

(*c*) For a concentration of 1 g/cm^3 = 10^3 kg/m^3 of an optically active substance of molar mass 100 g/mol = 0.1 kg/mol, determine the difference between the indices of refraction in the solution for left and right circularly polarized light. Compare this difference with their average value of about 1.40.

9. Calculate the potential ϕ at all points inside a cube of side h when there are no charges within the cube and when the potential on the side $x = 0$ has a constant value ϕ_0 while the potential on the other five sides is zero.

10. Define two new quantities $\mathbf{A}'$ and ϕ' in terms of the vector potential **A** and the scalar potential ϕ by

$$\mathbf{A}' = \mathbf{A} - \nabla\psi \qquad \phi' = \phi + \frac{\partial \psi}{\partial t}$$

where $\psi(x,y,z,t)$ is any scalar function of x, y, z, and t.

(*a*) Show that Maxwell's equations are invariant to this *gauge transformation.*

(*b*) If **A**, ϕ and $\mathbf{A}'$, ϕ' satisfy the Lorentz gauge, show that $\psi(x,y,z,t)$ obeys the wave equation (4-64).

11. Let the magnetic induction **B** be parallel to the z axis and have a constant magnitude B_0. Construct at least two vector potentials **A** which satisfy the Coulomb gauge and which yield **B**.

Chapter five

Dyadics

5-1 Definition of a dyadic

In some physical situations it is necessary to express each component of a vector **A** as a linear function of the components of another vector **B**,

$$\begin{aligned} A_x &= T_{xx}B_x + T_{xy}B_y + T_{xz}B_z \\ A_y &= T_{yx}B_x + T_{yy}B_y + T_{yz}B_z \\ A_z &= T_{zx}B_x + T_{zy}B_y + T_{zz}B_z \end{aligned} \tag{5-1}$$

where T_{xx}, etc., are coefficients. Thus we need a general mathematical method for relating the vector **A** to the vector **B**. In three-dimensional cartesian space, there exists an especially simple representation, known as the dyadic formalism, for such transformations. For many purposes it is sufficient to consider only dyadics, which are a special kind of second-rank tensor (often called just *tensor*).

In order to present the equations in this chapter in more compact form, we use numerical subscripts rather than letter subscripts for the vector components. Thus, we write

$$\mathbf{A} = \mathbf{i}A_1 + \mathbf{j}A_2 + \mathbf{k}A_3$$

$$\mathbf{B} = \mathbf{i}B_1 + \mathbf{j}B_2 + \mathbf{k}B_3$$

Equations (5-1) then become

$$\begin{aligned} A_1 &= T_{11}B_1 + T_{12}B_2 + T_{13}B_3 \\ A_2 &= T_{21}B_1 + T_{22}B_2 + T_{23}B_3 \\ A_3 &= T_{31}B_1 + T_{32}B_2 + T_{33}B_3 \end{aligned} \tag{5-2}$$

which can be expressed more compactly as

$$A_i = \sum_{j=1}^{3} T_{ij} B_j \qquad i = 1, 2, 3 \tag{5-3}$$

If we multiply together two vectors **A** and **B** without a dot or cross between them, we have

$$\begin{aligned} \mathbf{AB} = {} & \mathbf{ii}A_1B_1 + \mathbf{ij}A_1B_2 + \mathbf{ik}A_1B_3 \\ & + \mathbf{ji}A_2B_1 + \mathbf{jj}A_2B_2 + \mathbf{jk}A_2B_3 \\ & + \mathbf{ki}A_3B_1 + \mathbf{kj}A_3B_2 + \mathbf{kk}A_3B_3 \end{aligned} \tag{5-4}$$

The product has nine terms, and each term consists of a pair of unit vectors with a scalar coefficient. A pair of unit vectors is called a *unit dyad*, of which there are nine possible.

If we multiply each pair of unit dyads by a scalar coefficient T_{ij} and add, we have

$$\begin{aligned} \mathsf{T} = {} & \mathbf{ii}T_{11} + \mathbf{ij}T_{12} + \mathbf{ik}T_{13} \\ & + \mathbf{ji}T_{21} + \mathbf{jj}T_{22} + \mathbf{jk}T_{23} \\ & + \mathbf{ki}T_{31} + \mathbf{kj}T_{32} + \mathbf{kk}T_{33} \end{aligned} \tag{5-5}$$

Equation (5-5) shows a *dyadic* T in its most general representation. Dyadics are usually denoted by boldface *sans serif* type. In handwriting a dyadic may be denoted by a double wavy underscore or by a double arrow over the symbol. The quantities T_{ij} are called the *components* of the dyadic. The product of two vectors always gives a dyadic, but in general a dyadic cannot be expressed as the product of two vectors.

The sum of two dyadics R and S is another dyadic T, whose components are equal to the sums of the corresponding components of R and S,

$$\mathsf{T} = \mathsf{R} + \mathsf{S} \tag{5-6}$$

such that

$$T_{ij} = R_{ij} + S_{ij} \tag{5-7}$$

5-2 Multiplication by a dyadic

The product of a scalar α and a dyadic T is another dyadic S, whose components are α times the corresponding components of T,

$$\mathsf{S} = \alpha \mathsf{T} \tag{5-8}$$

such that

$$S_{ij} = \alpha T_{ij} \tag{5-9}$$

The dot product of a dyadic and a vector is another vector. If the dyadic can be written as the product of two vectors **A** and **B**, we have

$$\begin{aligned} (\mathbf{AB}) \cdot \mathbf{C} &= \mathbf{A}(\mathbf{B} \cdot \mathbf{C}) \\ \mathbf{C} \cdot (\mathbf{AB}) &= (\mathbf{C} \cdot \mathbf{A})\mathbf{B} \end{aligned} \tag{5-10}$$

If the dyadic cannot be written as the product of two vectors, then the dot product of a vector **A** and a dyadic **T** is

$$\begin{aligned} \mathbf{T} \cdot \mathbf{A} &= \mathbf{B} \\ \mathbf{A} \cdot \mathbf{T} &= \mathbf{C} \end{aligned} \tag{5-11}$$

where **B** and **C** are the resulting vectors. To find **B** or **C** we use the following relationships between the unit vectors and unit dyads:

$$\begin{aligned} (\mathbf{ii}) \cdot \mathbf{i} &= \mathbf{i}(\mathbf{i} \cdot \mathbf{i}) = \mathbf{i} & \mathbf{i} \cdot (\mathbf{ii}) &= (\mathbf{i} \cdot \mathbf{i})\mathbf{i} = \mathbf{i} \\ (\mathbf{ij}) \cdot \mathbf{i} &= \mathbf{i}(\mathbf{j} \cdot \mathbf{i}) = 0 & \mathbf{i} \cdot (\mathbf{ij}) &= (\mathbf{i} \cdot \mathbf{i})\mathbf{j} = \mathbf{j} \\ (\mathbf{ji}) \cdot \mathbf{i} &= \mathbf{j}(\mathbf{i} \cdot \mathbf{i}) = \mathbf{j} & \mathbf{i} \cdot (\mathbf{ji}) &= (\mathbf{i} \cdot \mathbf{j})\mathbf{i} = 0 \\ &\text{etc.} & &\text{etc.} \end{aligned}$$

Equations (5-11) may also be written in component form:

$$\begin{aligned} B_i &= \sum_{j=1}^{3} T_{ij} A_j \qquad i = 1, 2, 3 \\ C_i &= \sum_{j=1}^{3} A_j T_{ji} \qquad i = 1, 2, 3 \end{aligned} \tag{5-12}$$

In general, the commutative law of multiplication does not hold, so that

$$\mathbf{A} \cdot \mathbf{T} \neq \mathbf{T} \cdot \mathbf{A}$$

The dot product of two dyadics is a third dyadic. If both dyadics can be written as products of vectors, then

$$(\mathbf{AB}) \cdot (\mathbf{CD}) = \mathbf{A}(\mathbf{B} \cdot \mathbf{C})\mathbf{D} = (\mathbf{B} \cdot \mathbf{C})\mathbf{AD} \tag{5-13}$$

If only one dyadic can be written as a product of two vectors, then

$$\begin{aligned} (\mathbf{AB}) \cdot \mathbf{T} &= \mathbf{A}(\mathbf{B} \cdot \mathbf{T}) \\ \mathbf{S} \cdot (\mathbf{CD}) &= (\mathbf{S} \cdot \mathbf{C})\mathbf{D} \end{aligned} \tag{5-14}$$

If neither dyadic can be written as a product of two vectors, then

$$\mathbf{R} \cdot \mathbf{S} = \mathbf{T} \tag{5-15}$$

where **T** is the resulting dyadic. The components of **T** are given by

$$T_{ij} = \sum_{k=1}^{3} R_{ik} S_{kj} \tag{5-16}$$

The double dot product of a dyadic with two vectors is a scalar. If the dyadic can be written as the product of two vectors, then

$$\mathbf{D} \cdot (\mathbf{AB}) \cdot \mathbf{C} = (\mathbf{D} \cdot \mathbf{A})(\mathbf{B} \cdot \mathbf{C}) \tag{5-17}$$

If the dyadic cannot be written as the product of two vectors, then

$$\begin{aligned} \mathbf{A} \cdot \mathbf{T} \cdot \mathbf{B} &= (\mathbf{A} \cdot \mathbf{T}) \cdot \mathbf{B} = \mathbf{A} \cdot (\mathbf{T} \cdot \mathbf{B}) = \alpha \\ \mathbf{B} \cdot \mathbf{T} \cdot \mathbf{A} &= (\mathbf{B} \cdot \mathbf{T}) \cdot \mathbf{A} = \mathbf{B} \cdot (\mathbf{T} \cdot \mathbf{A}) = \beta \end{aligned} \tag{5-18}$$

where α and β are the resulting scalars. Equations (5-18) in component form are

$$\begin{aligned} \alpha &= \mathbf{A} \cdot \mathbf{T} \cdot \mathbf{B} = \sum_{i=1}^{3} \sum_{j=1}^{3} A_i B_j T_{ij} \\ \beta &= \mathbf{B} \cdot \mathbf{T} \cdot \mathbf{A} = \sum_{i=1}^{3} \sum_{j=1}^{3} A_j B_i T_{ij} \end{aligned} \tag{5-19}$$

A special case is the double dot product of two dyadics **S** and **T**,

$$\mathbf{S} : \mathbf{T} = \mathbf{T} : \mathbf{S} = \sum_{i=1}^{3} \sum_{j=1}^{3} S_{ij} T_{ji} \tag{5-20}$$

Some authors define

$$\mathbf{S} : \mathbf{T} = \sum_{i=1}^{3} \sum_{j=1}^{3} S_{ij} T_{ij}$$

The choice is merely one of convention. We shall select Eq. (5-20) as the definition of $\mathbf{S} : \mathbf{T}$ in this book. In terms of the double dot product defined by Eq. (5-20), we obtain

$$\mathbf{D} \cdot (\mathbf{AB}) \cdot \mathbf{C} = \mathbf{AB} : \mathbf{CD} = \mathbf{CD} : \mathbf{AB} \tag{5-21}$$

and Eqs. (5-18) become

$$\begin{aligned} \alpha &= \mathbf{A} \cdot \mathbf{T} \cdot \mathbf{B} = \mathbf{T} : \mathbf{BA} = \mathbf{BA} : \mathbf{T} \\ \beta &= \mathbf{B} \cdot \mathbf{T} \cdot \mathbf{A} = \mathbf{T} : \mathbf{AB} = \mathbf{AB} : \mathbf{T} \end{aligned} \tag{5-22}$$

The cross product of a dyadic with a vector is another dyadic. If the dyadic is the product of two vectors, we have

$$\begin{aligned} (\mathbf{AB}) \times \mathbf{C} &= \mathbf{A}(\mathbf{B} \times \mathbf{C}) \\ \mathbf{C} \times (\mathbf{AB}) &= (\mathbf{C} \times \mathbf{A})\mathbf{B} \end{aligned} \tag{5-23}$$

If the dyadic is not the product of two vectors, then we have

$$\mathbf{A} \times \mathbf{T} = \mathbf{S} \qquad (5\text{-}24)$$
$$\mathbf{T} \times \mathbf{A} = \mathbf{R}$$

where **R** and **S** are the resulting dyadics. To find **R** or **S** we use the following relationships, which are valid for a *right-handed* cartesian coordinate system:

$$(\mathbf{ii}) \times \mathbf{i} = \mathbf{i}(\mathbf{i} \times \mathbf{i}) = 0 \qquad \mathbf{i} \times (\mathbf{ii}) = (\mathbf{i} \times \mathbf{i})\mathbf{i} = 0$$
$$(\mathbf{ij}) \times \mathbf{i} = \mathbf{i}(\mathbf{j} \times \mathbf{i}) = -\mathbf{ik} \qquad \mathbf{i} \times (\mathbf{ij}) = (\mathbf{i} \times \mathbf{i})\mathbf{j} = 0$$
$$(\mathbf{ji}) \times \mathbf{i} = \mathbf{j}(\mathbf{i} \times \mathbf{i}) = 0 \qquad \mathbf{i} \times (\mathbf{ji}) = (\mathbf{i} \times \mathbf{j})\mathbf{i} = \mathbf{ki}$$

etc. etc.

5-3 Some examples

In order to illustrate the algebraic procedures described in the previous two sections, we consider the specific vectors and dyadics

$$\mathbf{A} = 2\mathbf{i} + \mathbf{j} - 2\mathbf{k} \qquad \mathbf{T} = 5\mathbf{ii} - \mathbf{ik} + 3\mathbf{jk} + 2\mathbf{ki}$$
$$\mathbf{B} = -2\mathbf{i} + 3\mathbf{k} \qquad \mathbf{S} = 2\mathbf{ii} + 3\mathbf{ik} - \mathbf{kj}$$

and calculate the following quantities:

(*a*) $\mathbf{T} + \mathbf{S}$ (*c*) $\mathbf{A} \cdot \mathbf{T}$ (*e*) $\mathbf{A} \cdot \mathbf{T} \cdot \mathbf{B}$ (*g*) $\mathbf{S} \cdot \mathbf{T}$ (*i*) $\mathbf{B} \times \mathbf{S}$

(*b*) $3\mathbf{T}$ (*d*) $\mathbf{T} \cdot \mathbf{A}$ (*f*) $\mathbf{B} \cdot \mathbf{T} \cdot \mathbf{A}$ (*h*) $\mathbf{S} : \mathbf{T}$

The solutions are:

(*a*) $\mathbf{T} + \mathbf{S} = 7\mathbf{ii} + 2\mathbf{ik} + 3\mathbf{jk} + 2\mathbf{ki} - \mathbf{kj}$

(*b*) $3\mathbf{T} = 15\mathbf{ii} - 3\mathbf{ik} + 9\mathbf{jk} + 6\mathbf{ki}$

(*c*) $\mathbf{A} \cdot \mathbf{T} = (2\mathbf{i} + \mathbf{j} - 2\mathbf{k}) \cdot (5\mathbf{ii} - \mathbf{ik} + 3\mathbf{jk} + 2\mathbf{ki})$
$= 10\mathbf{i} - 2\mathbf{k} + 3\mathbf{k} - 4\mathbf{i} = 6\mathbf{i} + \mathbf{k}$

(*d*) $\mathbf{T} \cdot \mathbf{A} = (5\mathbf{ii} - \mathbf{ik} + 3\mathbf{jk} + 2\mathbf{ki}) \cdot (2\mathbf{i} + \mathbf{j} - 2\mathbf{k})$
$= 10\mathbf{i} + 4\mathbf{k} + 2\mathbf{i} - 6\mathbf{j} = 12\mathbf{i} - 6\mathbf{j} + 4\mathbf{k}$

(*e*) $\mathbf{A} \cdot \mathbf{T} \cdot \mathbf{B} = (6\mathbf{i} + \mathbf{k}) \cdot (-2\mathbf{i} + 3\mathbf{k}) = -12 + 3 = -9$

(*f*) $\mathbf{B} \cdot \mathbf{T} \cdot \mathbf{A} = (-2\mathbf{i} + 3\mathbf{k}) \cdot (12\mathbf{i} - 6\mathbf{j} + 4\mathbf{k}) = -24 + 12 = -12$

(*g*) $\mathbf{S} \cdot \mathbf{T} = (2\mathbf{ii} + 3\mathbf{ik} - \mathbf{kj}) \cdot (5\mathbf{ii} - \mathbf{ik} + 3\mathbf{jk} + 2\mathbf{ki})$
$= 10\mathbf{ii} - 2\mathbf{ik} + 6\mathbf{ii} - 3\mathbf{kk} = 16\mathbf{ii} - 2\mathbf{ik} - 3\mathbf{kk}$

(*h*) $\mathbf{S} : \mathbf{T} = (2\mathbf{ii} + 3\mathbf{ik} - \mathbf{kj}) : (5\mathbf{ii} - \mathbf{ik} + 3\mathbf{jk} + 2\mathbf{ki})$
$= 10 + 6 - 3 = 13$

(*i*) $$\begin{aligned}\mathbf{B} \times \mathbf{S} &= (-2\mathbf{i} + 3\mathbf{k}) \times (2\mathbf{ii} + 3\mathbf{ik} - \mathbf{kj}) \\ &= 2(\mathbf{i} \times \mathbf{k})\mathbf{j} + 6(\mathbf{k} \times \mathbf{i})\mathbf{i} + 9(\mathbf{k} \times \mathbf{i})\mathbf{k} \\ &= -2\mathbf{jj} + 6\mathbf{ji} + 9\mathbf{jk}\end{aligned}$$

5-4 Dyadics with differential operators

In the above discussion the vector operator ∇,

$$\nabla = \mathbf{i}\frac{\partial}{\partial x} + \mathbf{j}\frac{\partial}{\partial y} + \mathbf{k}\frac{\partial}{\partial z}$$

may be used in place of a vector without changing the validity of the results. In particular, we may have the dyadic operator $\nabla\nabla$,

$$\begin{aligned}\nabla\nabla &= \mathbf{ii}\frac{\partial^2}{\partial x^2} + \mathbf{ij}\frac{\partial}{\partial x}\frac{\partial}{\partial y} + \mathbf{ik}\frac{\partial}{\partial x}\frac{\partial}{\partial z} \\ &+ \mathbf{ji}\frac{\partial}{\partial y}\frac{\partial}{\partial x} + \mathbf{jj}\frac{\partial^2}{\partial y^2} + \mathbf{jk}\frac{\partial}{\partial y}\frac{\partial}{\partial z} \\ &+ \mathbf{ki}\frac{\partial}{\partial z}\frac{\partial}{\partial x} + \mathbf{kj}\frac{\partial}{\partial z}\frac{\partial}{\partial y} + \mathbf{kk}\frac{\partial^2}{\partial z^2}\end{aligned} \tag{5-25}$$

The dyadic $\nabla\mathbf{A}$ is

$$\begin{aligned}\nabla\mathbf{A} &= \mathbf{ii}\frac{\partial}{\partial x}A_1 + \mathbf{ij}\frac{\partial}{\partial x}A_2 + \mathbf{ik}\frac{\partial}{\partial x}A_3 \\ &+ \mathbf{ji}\frac{\partial}{\partial y}A_1 + \mathbf{jj}\frac{\partial}{\partial y}A_2 + \mathbf{jk}\frac{\partial}{\partial y}A_3 \\ &+ \mathbf{ki}\frac{\partial}{\partial z}A_1 + \mathbf{kj}\frac{\partial}{\partial z}A_2 + \mathbf{kk}\frac{\partial}{\partial z}A_3\end{aligned} \tag{5-26}$$

while the dyadic $\mathbf{A}\nabla$ is a dyadic operator,

$$\begin{aligned}\mathbf{A}\nabla &= \mathbf{ii}\,A_1\frac{\partial}{\partial x} + \mathbf{ij}\,A_1\frac{\partial}{\partial y} + \mathbf{ik}\,A_1\frac{\partial}{\partial z} \\ &+ \mathbf{ji}\,A_2\frac{\partial}{\partial x} + \mathbf{jj}\,A_2\frac{\partial}{\partial y} + \mathbf{jk}\,A_2\frac{\partial}{\partial z} \\ &+ \mathbf{ki}\,A_3\frac{\partial}{\partial x} + \mathbf{kj}\,A_3\frac{\partial}{\partial y} + \mathbf{kk}\,A_3\frac{\partial}{\partial z}\end{aligned} \tag{5-27}$$

In Eqs. (5-11), if **A** is replaced by ∇, we have

$$\mathbf{T}\cdot\nabla = \mathbf{i}\sum_{i=1}^{3} T_{1i}\frac{\partial}{\partial x_i} + \mathbf{j}\sum_{i=1}^{3} T_{2i}\frac{\partial}{\partial x_i} + \mathbf{k}\sum_{i=1}^{3} T_{3i}\frac{\partial}{\partial x_i} \tag{5-28}$$

$$\nabla\cdot\mathbf{T} = \mathbf{i}\sum_{i=1}^{3}\frac{\partial}{\partial x_i}T_{i1} + \mathbf{j}\sum_{i=1}^{3}\frac{\partial}{\partial x_i}T_{i2} + \mathbf{k}\sum_{i=1}^{3}\frac{\partial}{\partial x_i}T_{i3} \tag{5-29}$$

where the notation x_1, x_2, x_3 for x, y, z, respectively, has been introduced for compactness. The quantity $\mathbf{T} \cdot \nabla$ is a vector operator and $\nabla \cdot \mathbf{T}$, the divergence of a dyadic, is a vector. The divergence of a dyadic which can be written as the product of two vectors may be expanded as follows: the ith component of $\nabla \cdot (\mathbf{AB})$ is

$$[\nabla \cdot (\mathbf{AB})]_i = \sum_{j=1}^{3} \frac{\partial}{\partial x_j} A_j B_i = \sum_{j=1}^{3} A_j \frac{\partial B_i}{\partial x_j} + \sum_{j=1}^{3} B_i \frac{\partial A_j}{\partial x_j}$$
$$= (\mathbf{A} \cdot \nabla) B_i + B_i \nabla \cdot \mathbf{A} \qquad i = 1, 2, 3$$

Therefore, we have

$$\nabla \cdot (\mathbf{AB}) = (\mathbf{A} \cdot \nabla)\mathbf{B} + \mathbf{B} \nabla \cdot \mathbf{A} \tag{5-30}$$

5-5 Unit dyadic

The unit dyadic is defined as

$$\mathbf{1} = \mathbf{ii} + \mathbf{jj} + \mathbf{kk} \tag{5-31}$$

and has the property that

$$\begin{aligned} \mathbf{A} \cdot \mathbf{1} &= \mathbf{A} \\ \mathbf{1} \cdot \mathbf{A} &= \mathbf{A} \end{aligned} \tag{5-32}$$

The inverse dyadic $\mathbf{T}^{-1}$ of the dyadic $\mathbf{T}$ is defined by the relation

$$\mathbf{T} \cdot \mathbf{T}^{-1} = \mathbf{1} \tag{5-33}$$

With the use of the unit dyadic, the dot or scalar product of two vectors may be expressed as the double dot product of two dyadics,

$$\mathbf{A} \cdot \mathbf{B} = \sum_{i=1}^{3} A_i B_i = \sum_{i=1}^{3} \sum_{j=1}^{3} \delta_{ij} A_j B_i = \mathbf{1} : (\mathbf{AB}) \tag{5-34}$$

where δ_{ij} is the Kronecker delta function,

$$\begin{aligned} \delta_{ij} &= 1 \qquad i = j \\ \delta_{ij} &= 0 \qquad i \neq j \end{aligned} \tag{5-35}$$

Similarly, if $\mathbf{A}$ is replaced by the operator ∇, we have

$$\nabla \cdot \mathbf{B} = \sum_{i=1}^{3} \frac{\partial B_i}{\partial x_i} = \sum_{i=1}^{3} \sum_{j=1}^{3} \delta_{ij} \frac{\partial B_i}{\partial x_j} = \mathbf{1} : (\nabla \mathbf{B}) \tag{5-36}$$

The gradient of a scalar α may be expressed as the divergence of a dyadic,

$$\nabla \alpha = \nabla \cdot \alpha \mathbf{1} \tag{5-37}$$

5-6 Trace of a dyadic

The trace of a dyadic $\mathbf{T}$ is the sum of the diagonal components T_{11}, T_{22}, T_{33},

$$\mathrm{Tr}\,\mathbf{T} = T_{11} + T_{22} + T_{33} \tag{5-38}$$

and is therefore a scalar. The trace of the dyadic $\nabla\mathbf{A}$, for example, is

$$\mathrm{Tr}\,\nabla\mathbf{A} = \sum_{i=1}^{3} \frac{\partial A_i}{\partial x_i} = \nabla \cdot \mathbf{A} \tag{5-39}$$

5-7 Symmetric and antisymmetric dyadics

A dyadic $\mathbf{T}$ is symmetric if $T_{ij} = T_{ji}$. For a symmetric dyadic $\mathbf{T}_s$, the commutative law of multiplication holds:

$$\begin{aligned} \mathbf{T}_s \cdot \mathbf{A} &= \mathbf{A} \cdot \mathbf{T}_s \\ \mathbf{A} \cdot \mathbf{T}_s \cdot \mathbf{B} &= \mathbf{B} \cdot \mathbf{T}_s \cdot \mathbf{A} \end{aligned} \tag{5-40}$$

A dyadic of the form $\mathbf{AA}$ is symmetric. A dyadic is antisymmetric if $T_{ij} = -T_{ji}$ for $i \neq j$ and $T_{ii} = 0$. For an antisymmetric dyadic $\mathbf{T}_a$,

$$\mathbf{A} \cdot \mathbf{T}_a = -\mathbf{T}_a \cdot \mathbf{A} \tag{5-41}$$

It is always possible to express an arbitrary asymmetric dyadic as the sum of a symmetric and an antisymmetric dyadic. Each component of an asymmetric dyadic $\mathbf{T}$ may be written

$$T_{ij} = \frac{1}{2}(T_{ij} + T_{ji}) + \frac{1}{2}(T_{ij} - T_{ji}) \tag{5-42}$$

If we define a symmetric dyadic $\mathbf{T}_s$ with components $\frac{1}{2}(T_{ij} + T_{ji})$ and an antisymmetric dyadic $\mathbf{T}_a$ with components $\frac{1}{2}(T_{ij} - T_{ji})$, then the dyadic $\mathbf{T}$ may be written in the form

$$\mathbf{T} = \mathbf{T}_s + \mathbf{T}_a \tag{5-43}$$

and is therefore expressed as a sum of a symmetric and an antisymmetric part.

The symmetric or antisymmetric character of a dyadic is invariant to a rotation of the coordinate system. For a symmetric dyadic it is always possible to rotate the coordinate axes to a position such that the off-diagonal components T_{ij} $(i \neq j)$ of the dyadic vanish and $\mathbf{T}_s = \mathbf{ii}T_{11} + \mathbf{jj}T_{22} + \mathbf{kk}T_{33}$. These coordinate axes are called the principal axes of the dyadic and the dyadic is then said to be diagonal.

5-8 Vector product

The vector product of two vectors **B** and **C** is a mathematical operation which produces a new vector **A**. It is possible, therefore, to express such an operation as a transformation involving dyadics. Thus, the three relations

$$\mathbf{A} = \mathbf{B} \times \mathbf{C} \tag{5-44}$$

$$\mathbf{A} = \mathbf{S} \cdot \mathbf{C} \tag{5-45}$$

$$\mathbf{A} = \mathbf{B} \cdot \mathbf{T} \tag{5-46}$$

are equivalent and the problem is to determine the dyadics **S** and **T**.

In the expansion of Eq. (5-44) we have

$$\begin{aligned} A_x &= B_y C_z - B_z C_y \\ A_y &= B_z C_x - B_x C_z \\ A_z &= B_x C_y - B_y C_x \end{aligned} \tag{5-47}$$

Writing out Eq. (5-45) we find

$$\begin{aligned} A_x &= S_{xx} C_x + S_{xy} C_y + S_{xz} C_z \\ A_y &= S_{yx} C_x + S_{yy} C_y + S_{yz} C_z \\ A_z &= S_{zx} C_x + S_{zy} C_y + S_{zz} C_z \end{aligned} \tag{5-48}$$

Comparing Eqs. (5-47) with (5-48) term by term, we see that

$$\begin{aligned} S_{xx} &= S_{yy} = S_{zz} = 0 \\ S_{yx} &= -S_{xy} = B_z \\ S_{xz} &= -S_{zx} = B_y \\ S_{zy} &= -S_{yz} = B_x \end{aligned} \tag{5-49}$$

so that

$$\mathbf{S} = -\mathbf{ij}B_z + \mathbf{ik}B_y + \mathbf{ji}B_z - \mathbf{jk}B_x - \mathbf{ki}B_y + \mathbf{kj}B_x \tag{5-50}$$

Similarly, expansion of Eq. (5-46) and term-by-term comparison with Eqs. (5-47) give

$$\mathbf{T} = -\mathbf{ij}C_z + \mathbf{ik}C_y + \mathbf{ji}C_z - \mathbf{jk}C_x - \mathbf{ki}C_y + \mathbf{kj}C_x \tag{5-51}$$

Note that both **S** and **T** are antisymmetric dyadics.

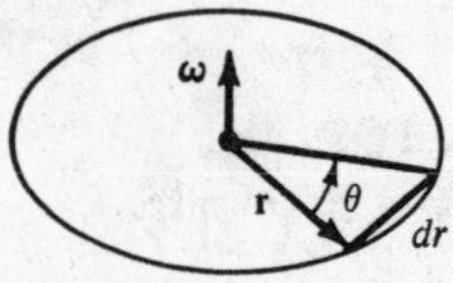

FIGURE 5-1

5-9 Moment of inertia of a rigid body

A rigid body is one in which all the distances between the component particles remain fixed with time. However, the body as a whole can move through space (translational motion) and rotate about any axis in space (rotational motion). In order to separate the translational motion from the rotational motion, we take the center of mass of the system of particles as the origin of a cartesian coordinate system and refer the rotational motion to an axis passing through this origin.

Consider a single particle moving in a circular orbit about the origin at a fixed distance r. Assume that the angular velocity $\omega = d\theta/dt$, where θ is the angle shown in Fig. 5-1, is constant with time. The magnitude of the velocity is, then,

$$v = r\frac{d\theta}{dt} = \omega r \tag{5-52}$$

If $\boldsymbol{\omega}$ is a vector of magnitude ω perpendicular to the plane of motion, the velocity vector $\mathbf{v}$ is

$$\mathbf{v} = \boldsymbol{\omega} \times \mathbf{r} \tag{5-53}$$

In classical mechanics the *angular momentum* $\mathbf{L}$ of a single particle referred to an arbitrary fixed point as origin is defined as

$$\mathbf{L} = \mathbf{r} \times \mathbf{p} = m(\mathbf{r} \times \mathbf{v}) \tag{5-54}$$

where $\mathbf{p}$ is the linear momentum $(= m\mathbf{v})$ of the particle and m is its mass. For a collection of particles with mass m_i, the total angular momentum is

$$\mathbf{L} = \Sigma_i m_i(\mathbf{r}_i \times \mathbf{v}_i) \tag{5-55}$$

In a rigid body each particle i has the same angular velocity $\boldsymbol{\omega}$, and so we may write

$$\mathbf{v}_i = \boldsymbol{\omega} \times \mathbf{r}_i \tag{5-56}$$

and

$$\mathbf{L} = \Sigma_i m_i \mathbf{r}_i \times (\boldsymbol{\omega} \times \mathbf{r}_i) \tag{5-57}$$

At this point in our discussion we may follow two lines of development, which are, however, equivalent. In the first, we use Eq. (1-33) to write Eq. (5-57) in the form

$$\begin{aligned}\mathbf{L} &= \Sigma_i m_i [r_i^2\boldsymbol{\omega} - \mathbf{r}_i(\mathbf{r}_i \cdot \boldsymbol{\omega})] \\ &= \Sigma_i m_i(r_i^2 \mathbf{1} - \mathbf{r}_i\mathbf{r}_i) \cdot \boldsymbol{\omega} \\ &= \mathbf{I} \cdot \boldsymbol{\omega} \end{aligned} \tag{5-58}$$

where the *moment of inertia* dyadic is

$$\begin{aligned}\mathbf{I} &= \sum_i m_i(r_i^2 \mathbf{1} - \mathbf{r}_i\mathbf{r}_i) \\ &= \sum_i m_i [\mathbf{ii}(y_i^2 + z_i^2) - \mathbf{ij}x_iy_i - \mathbf{ik}x_iz_i - \mathbf{ji}x_iy_i + \mathbf{jj}(x_i^2 + z_i^2) \\ &\qquad - \mathbf{jk}y_iz_i - \mathbf{ki}x_iz_i - \mathbf{kj}y_iz_i + \mathbf{kk}(x_i^2 + y_i^2)] \end{aligned} \tag{5-59}$$

An alternative approach is to write Eq. (5-57) in the form of Eq. (5-45),

$$\mathbf{L} = \sum_i m_i \mathbf{T}_i \cdot (\boldsymbol{\omega} \times \mathbf{r}_i) \tag{5-60}$$

where

$$\mathbf{T}_i = -\mathbf{ij}z_i + \mathbf{ik}y_i + \mathbf{ji}z_i - \mathbf{jk}x_i - \mathbf{ki}y_i + \mathbf{kj}x_i \tag{5-61}$$

The vector $\mathbf{v}_i = \boldsymbol{\omega} \times \mathbf{r}_i$ may also be written in the form of Eq. (5-45),

$$\mathbf{v}_i = -\mathbf{r}_i \times \boldsymbol{\omega} = -\mathbf{S}_i \cdot \boldsymbol{\omega} \tag{5-62}$$

where the dyadic $\mathbf{S}_i$ happens to be the same as $\mathbf{T}_i$. Consequently, Eq. (5-60) takes the form

$$\mathbf{L} = -\sum_i m_i \mathbf{T}_i \cdot \mathbf{T}_i \cdot \boldsymbol{\omega} = \mathbf{I} \cdot \boldsymbol{\omega} \tag{5-63}$$

where the moment of inertia $\mathbf{I}$ is

$$\mathbf{I} = -\sum_i m_i \mathbf{T}_i \cdot \mathbf{T}_i \tag{5-64}$$

When $\mathbf{I}$ is computed according to Eq. (5-64) with $\mathbf{T}_i$ in Eq. (5-61), the expression (5-59) is obtained.

The moment of inertia $\mathbf{I}$ in Eq. (5-59) is a symmetric dyadic. Therefore, the cartesian coordinate axes x, y, z can be rotated while retaining the center of mass as origin so that the elements I_{ij} for $i \neq j$ vanish. The moment of inertia then takes the form

$$\mathbf{I} = \mathbf{ii}I_{11} + \mathbf{jj}I_{22} + \mathbf{kk}I_{33} \tag{5-65}$$

The new set of coordinate axes are called the *principal axes* and the components I_{11}, I_{22}, I_{33} are called the *principal moments of inertia.*

PROBLEMS

1. The vectors **A** and **B** and the dyadic **T** are given by

$$\mathbf{A} = \mathbf{i} + 2\mathbf{j} - \mathbf{k}$$
$$\mathbf{B} = 3\mathbf{i} - \mathbf{j}$$
$$\mathbf{T} = 2\mathbf{ii} - \mathbf{ij} + \mathbf{jk}$$

Find:

(*a*) $\mathbf{AB}$ (*b*) $\mathbf{AB} + \mathbf{T}$
(*c*) $\mathbf{T} \cdot \mathbf{A}$ (*d*) $\mathbf{A} \cdot \mathbf{T}$
(*e*) $\mathbf{T} \times \mathbf{A}$ (*f*) $\mathbf{A} \cdot \mathbf{T} \cdot \mathbf{B}$
(*g*) $\mathbf{T} \cdot \mathbf{AB}$ (*h*) $\mathbf{T} : \mathbf{AB}$
(*i*) $\mathrm{Tr}\,(\mathbf{AB})$

2. Express **AB** in Prob. 1 as the sum of a symmetric and an antisymmetric dyadic.

3. If **A**, **B**, **C**, and **D** are any vectors, show that $\mathbf{AB}:\mathbf{CD} = \mathbf{AC}:\mathbf{BD}$.

4. For the vector field **A** given by

$$\mathbf{A} = \mathbf{i}x^2 + \mathbf{j}y^2 + \mathbf{k}z^2$$

expand $\mathbf{A} \cdot \nabla\nabla \cdot \mathbf{A}$.

5. If **A** is any vector and **T** is any dyadic, express the following quantities in summation notation involving the components of **A** and **T**:

(*a*) $\mathbf{T} : \nabla\mathbf{A}$
(*b*) $\nabla \cdot \mathbf{T} \cdot \mathbf{A}$
(*c*) $\mathbf{A} \cdot \nabla \cdot \mathbf{T}$

6. Show that $\nabla\mathbf{r} = \mathbf{1}$, where $\mathbf{r} = \mathbf{i}x + \mathbf{j}y + \mathbf{k}z$.

7. Show that the components of the dyadic **T** in Eq. (5-46) are those given in Eq. (5-51).

8. Show that the dyadic $\mathbf{S}_i$ in Eq. (5-62) is equal to the dyadic $\mathbf{T}_i$ in Eq. (5-61).

9. The kinetic energy E_K for a system of particles is

$$E_K = \frac{1}{2}\sum_i m_i v_i^2$$

Show from Eqs. (1-30), (5-56), and (5-57) that

$$E_K = \frac{1}{2}\boldsymbol{\omega} \cdot \mathbf{L}$$

10. The bond angle for the water molecule is 104.5° and the bond lengths are 0.975 Å. Find the principal axes and calculate the principal moments of inertia.

Chapter six

Fluid Mechanics and Thermodynamics

6.1 Fluid flow

In this chapter we are concerned with fluid systems which are in motion with respect to an external, laboratory-fixed set of coordinate axes. These fluid systems may be either liquid or gaseous mixtures with ν chemical components denoted by the running index $\alpha(\alpha = 1, 2, \ldots, \nu)$.

A volume element dv of the fluid is located by the position vector $\mathbf{r}$ in the external coordinate system. The composition of this volume element dv at a time t may be given in terms of the mass fractions x_α of the ν chemical components and the mass density ρ of dv. Alternatively, the composition of the volume element may be specified by the partial mass densities ρ_α of the ν components. The mass fractions and the partial mass densities for dv at time t are related by

$$x_\alpha = \frac{\rho_\alpha}{\rho} \qquad \alpha = 1, 2, \ldots, \nu \tag{6-1}$$

$$\sum_{\alpha=1}^{\nu} \rho_\alpha = \rho \tag{6-2}$$

$$\sum_{\alpha=1}^{\nu} x_\alpha = 1 \tag{6-3}$$

Equation (6-1) is simply the definition of the mass fraction in terms of the partial mass density of component α. Equation (6-2) expresses the conservation of mass in dv at time t, and Eq. (6-3) follows from Eqs. (6-1) and (6-2).

For a one-phase system in equilibrium, the partial mass densities ρ_α are uniform throughout the fluid and are also constant with time. For a nonequilibrium system, however, the ρ_α depend on the position $\mathbf{r}$ in space and on the time t, hence $\rho_\alpha(\mathbf{r}, t)$. Actually $\rho_\alpha(\mathbf{r}, t)$ are averages over microscopically large but macroscopically small cells or volume elements in space. Thus, ρ_α is the average mass density of the molecules of component α in a given small region. For mathematical simplicity the partial mass densities ρ_α and the thermodynamic functions to be introduced later in this chapter are considered to be defined at every point in space and to be everywhere continuous and differentiable.

The mass flow in a fluid may be described in terms of the local mean velocity $\mathbf{u}_\alpha$ for each component α. The velocity $\mathbf{u}_\alpha$ does not refer to the velocity of a particular molecule of α; rather, it refers to the average velocity with respect to the external coordinate system of the molecules of α in a microscopically large, macroscopically small region of the fluid. Thus, the local mean velocity $\mathbf{u}_\alpha$ is a field quantity and is a function of position and time, $\mathbf{u}_\alpha(\mathbf{r}, t)$.

The velocity $\mathbf{u}$ of the local center of mass may be defined by the relation

$$\rho \mathbf{u} = \sum_{\alpha=1}^{\nu} \rho_\alpha \mathbf{u}_\alpha \tag{6-4}$$

This velocity $\mathbf{u}$ varies, in general, with position $\mathbf{r}$ and time t. A coordinate system fixed relative to the local center of mass in one part of the fluid system usually moves with respect to a coordinate system fixed relative to the local center of mass in another part of the fluid system.

The mass current density or mass flux $\mathbf{J}_\alpha$ of component α relative to the external coordinate axes is

$$\mathbf{J}_\alpha = \rho_\alpha \mathbf{u}_\alpha \qquad \alpha = 1, 2, \ldots, \nu \tag{6-5}$$

This flux $\mathbf{J}_\alpha$ is the number of grams of component α crossing a plane of unit area in unit time when the plane is fixed relative to the external coordinate system and is oriented normal to the local mean velocity $\mathbf{u}_\alpha$. If this plane is oriented randomly with respect to the local mean velocity $\mathbf{u}_\alpha$, then the mass of α crossing the plane per unit area and time is $\mathbf{n} \cdot \mathbf{J}_\alpha$, where $\mathbf{n}$ is a unit vector normal to the plane. The units of $\mathbf{J}_\alpha$ are in grams of α per square centimeter per second.

The diffusion current density $\mathbf{j}_\alpha$ of component α relative to the local center of mass is defined as

$$\mathbf{j}_\alpha = \rho_\alpha(\mathbf{u}_\alpha - \mathbf{u}) \qquad \alpha = 1, 2, \ldots, \nu \tag{6-6}$$

The mass current density $\mathbf{J}_\alpha$ is related to the diffusion current density $\mathbf{j}_\alpha$ through Eqs. (6-5) and (6-6) by

$$\mathbf{J}_\alpha = \mathbf{j}_\alpha + \rho_\alpha \mathbf{u} \qquad \alpha = 1, 2, \ldots, \nu \tag{6-7}$$

Thus, the motion of material in a volume element dv is divided into convective motion described by $\mathbf{u}$, the velocity of the local center of mass, and diffusive motion of the ν components described by the $\mathbf{j}_\alpha$'s. The $\mathbf{j}_\alpha$'s are not all independent, for their sum vanishes:

$$\sum_{\alpha=1}^{\nu} \mathbf{j}_\alpha = \sum_{\alpha=1}^{\nu} \rho_\alpha(\mathbf{u}_\alpha - \mathbf{u}) = \rho\mathbf{u} - \rho\mathbf{u} = 0 \tag{6-8}$$

The substantial derivative d/dt of a field quantity measures the rate of change of the field quantity with respect to a coordinate system moving with the local center of mass. The substantial derivative is taken at the center of mass of the volume element dv, which is moving with velocity $\mathbf{u}$ with respect to the external coordinate axes. In this chapter the partial derivative notation $\partial/\partial t$ is used to indicate the time derivative taken at a point fixed with respect to the external coordinate axes and not moving with the fluid. The substantial derivative is related to the time derivative at a fixed point by

$$\begin{aligned}\frac{d}{dt} &= \frac{\partial}{\partial t} + \frac{dx}{dt}\frac{\partial}{\partial x} + \frac{dy}{dt}\frac{\partial}{\partial y} + \frac{dz}{dt}\frac{\partial}{\partial z} \\ &= \frac{\partial}{\partial t} + \mathbf{u} \cdot \nabla\end{aligned} \tag{6-9}$$

6-2 General conservation equation

We consider any extensive property $\mathcal{G}$ of the system, for example, the energy, the mass, a component of the momentum, etc. Associated with the property $\mathcal{G}$, is a specific quantity G, defined as $\mathcal{G}$ per unit mass. For a system which is not in equilibrium, G is not constant but is a function of position $\mathbf{r}$ and of time t. In an arbitrary volume V, fixed with respect to the external coordinate axes, the total quantity of $\mathcal{G}$ is

$$\mathcal{G} = \int_V \rho G \, dv \tag{6-10}$$

The time rate of change of $\mathcal{G}$ is given by applying Eq. (6-9) to Eq. (6-10). Since the volume V is fixed with respect to the external coordinate axes, the order of differentiation and integration may be interchanged, so that we have

$$\frac{dG}{dt} = \int_V \frac{\partial}{\partial t}(\rho G)\, dv \tag{6-11}$$

Another expression for dG/dt may be obtained as follows: The surface of the volume V is divided into infinitesimal elements of area ds. If $\mathbf{J}_G$ is the current density or flux of G, then the increase in G within V due to the flow of G through the boundary of V is just the negative of the surface integral of $\mathbf{J}_G$, $-\oint_S \mathbf{J}_G \cdot d\mathbf{s}$, where the integration is taken over the entire surface of V. The minus sign is necessary because a positive $d\mathbf{s}$ points outward so that a $\mathbf{J}_G$ parallel to $d\mathbf{s}$ results in a *decrease* in G within V. In the special case where G is carried only by convection due to actual flow of the medium, the current density $\mathbf{J}_G$ is given by

$$\mathbf{J}_G = \rho G \mathbf{u} \tag{6-12}$$

If ϕ_G is the internal source of G per unit volume and time (due to chemical reaction, for example), then the change G within V due to internal production is $\int_V \phi_G\, dv$ Adding the changes in G within V due to flow of G through the boundary of V and to the internal production of G, we have

$$\frac{dG}{dt} = -\oint_S \mathbf{J}_G \cdot d\mathbf{s} + \int_V \phi_G\, dv \tag{6-13}$$

The first integral on the right-hand side of Eq. (6-13) may be transformed by means of Gauss' theorem (3-61):

$$\oint_S \mathbf{J}_G \cdot d\mathbf{s} = \int_V \nabla \cdot \mathbf{J}_G\, dv \tag{6-14}$$

Combining Eqs. (6-11), (6-13), and (6-14), we find

$$\int_V \left[\frac{\partial(\rho G)}{\partial t} + \nabla \cdot \mathbf{J}_G - \phi_G\right] dv = 0 \tag{6-15}$$

Since the volume V under consideration is arbitrary, the only way for Eq. (6-15) to be valid for all volumes V is for the integrand to vanish. We therefore obtain a general equation of conservation:

$$\frac{\partial(\rho G)}{\partial t} + \nabla \cdot \mathbf{J}_G - \phi_G = 0 \tag{6-16}$$

6-3 Equation of continuity of matter

We now apply Eq. (6-16) to obtain the continuity equation for component α in the fluid system. This equation expresses the conservation of mass of component α. If G represents the total mass of α in the system, then G is the mass of α per unit mass of the system and equals ρ_α/ρ or x_α. The current density of mass of component α is the mass of α flowing per unit area in unit time, so that

$$\mathbf{J}_G = \rho G \mathbf{u}_\alpha = \rho_\alpha \mathbf{u}_\alpha = \mathbf{J}_\alpha \tag{6-17}$$

The internal source ϕ_G of mass of component α is the amount of α produced per unit volume and unit time by chemical reaction. We denote this quantity by ϕ_α. Substitution of these various values into Eq. (6-16) yields the equation of continuity for component α:

$$\frac{\partial \rho_\alpha}{\partial t} + \nabla \cdot (\rho_\alpha \mathbf{u}_\alpha) = \phi_\alpha \tag{6-18}$$

The equation of continuity for the total mass of the system may be obtained by summing Eq. (6-18) over all components. Since chemical reactions change some components into others but do not alter the total mass of the system, we have the relation

$$\sum_{\alpha=1}^{\nu} \phi_\alpha = 0 \tag{6-19}$$

The result of summing Eq. (6-18) is then

$$\frac{\partial \rho}{\partial t} + \nabla \cdot (\rho \mathbf{u}) = 0 \tag{6-20}$$

where Eqs. (6-2) and (6-4) have also been used.

Equation (6-20) can also be readily obtained from the general conservation equation (6-16) by setting $G = 1$. In this case the source term ϕ_G must vanish in order that the total mass of the system be conserved. From Eq. (6-12) the mass current density $\mathbf{J}_G$ is simply $\rho\mathbf{u}$.

The equation of continuity (6-20) may be used to express the general conservation equation (6-16) in terms of the substantial derivative. Expanding the derivative of the product $\partial(\rho G)/\partial t$ and applying Eq. (6-9) to G, we obtain

$$\rho \frac{dG}{dt} - \rho \mathbf{u} \cdot \nabla G + G \frac{\partial \rho}{\partial t} + \nabla \cdot \mathbf{J}_G = \phi_G \tag{6-21}$$

Since

$$\nabla \cdot (\rho G \mathbf{u}) = \rho \mathbf{u} \cdot \nabla G + G \nabla \cdot (\rho \mathbf{u}) \tag{6-22}$$

Eq. (6-21) becomes

$$\rho \frac{dG}{dt} + \nabla \cdot (\mathbf{J}_G - \rho G \mathbf{u}) + G\left[\frac{\partial \rho}{\partial t} + \nabla \cdot (\rho \mathbf{u})\right] = \phi_G \tag{6-23}$$

The quantity in brackets vanishes by Eq. (6-20) and the general conservation equation in terms of the substantial derivative is

$$\rho \frac{dG}{dt} + \nabla \cdot (\mathbf{J}_G - \rho G\mathbf{u}) = \phi_G \tag{6-24}$$

The equation of continuity for component α may be obtained in terms of the substantial derivative by setting $G = x_\alpha$ in Eq. (6-24),

$$\rho \frac{dx_\alpha}{dt} + \nabla \cdot \mathbf{j}_\alpha = \phi_\alpha \tag{6-25}$$

where Eqs. (6-7) and (6-17) are used. In order to obtain the equation of continuity for the total mass in terms of the substantial derivative, we apply Eq. (6-9) to Eq. (6-20) and obtain

$$\frac{d\rho}{dt} + \rho\nabla \cdot \mathbf{u} = 0 \tag{6-26}$$

In many cases, particularly for liquids, the mass density ρ is uniform throughout the fluid and is constant with time; i.e., there is no expansion or compression of the fluid. For such an *incompressible fluid* both $\partial\rho/\partial t$ and $\nabla\rho$ vanish and the equation of continuity (6-20) becomes

$$\nabla \cdot \mathbf{u} = 0 \tag{6-27}$$

6-4 Stress tensor

The surface force acting on the surface of a volume element V of fluid is the sum of the surface forces acting on each element of area ds and is

$$\int_S \boldsymbol{\sigma} \cdot d\mathbf{s}$$

where $\boldsymbol{\sigma}$ is a second-order tensor, or dyadic, known as the *stress tensor.* The component σ_{11} or σ_{xx} is the force per unit area in the x direction exerted on a plane surface which is perpendicular to the x direction. The quantities σ_{12} and σ_{13} (σ_{xy} and σ_{xz}) are the x components of the force per unit area exerted on plane surfaces normal to the y and z directions, respectively. Similar physical interpretations exist for the y and z components of the surface force. The components σ_{11}, σ_{22}, σ_{33} are called, therefore, normal stresses. The tangential elements σ_{12}, σ_{13}, σ_{21}, σ_{23}, σ_{31}, σ_{32} are called shear stresses. Some authors prefer to work with the negative of the stress tensor, a quantity called the pressure tensor, $\boldsymbol{\Pi} = -\boldsymbol{\sigma}$.

For a volume element of fluid which is in static equilibrium or which is moving with a constant velocity (i.e., is in static equilibrium in another galilean frame of reference), the total force and the total torque acting on the volume element vanish. From these relations the stress tensor $\boldsymbol{\sigma}$ may be shown to be symmetric. The stress tensor $\boldsymbol{\sigma}$ is symmetric, however, even for a fluid in which the local mean velocity **u** varies from point to point. This symmetry property may be shown generally from statistical mechanics for molecules that interact with central forces.

For a fluid with no internal friction, the surface force per unit area is just the external pressure p acting on the volume element V of fluid. Since the fluid is isotropic, the surface force is the same in all directions so that

$$\begin{aligned} &\sigma_{11} = \sigma_{22} = \sigma_{33} = -p \\ &\sigma_{12} = \sigma_{13} = \sigma_{21} = \sigma_{23} = \sigma_{31} = \sigma_{32} = 0 \end{aligned} \tag{6-28}$$

or

$$\boldsymbol{\sigma} = -p\mathbf{1} \tag{6-29}$$

Such a fluid is said to be an *ideal fluid.*

For a *viscous fluid* internal friction requires that an additional contribution to the stress tensor beyond that of the external pressure be considered. Thus we write the stress tensor $\boldsymbol{\sigma}$ for a viscous fluid in the form

$$\boldsymbol{\sigma} = -p\mathbf{1} + \boldsymbol{\sigma}' \tag{6-30}$$

where $\boldsymbol{\sigma}'$ is called the *viscosity stress tensor.* The internal friction occurs when the various volume elements of the fluid move with different velocities. The viscosity stress tensor $\boldsymbol{\sigma}'$ depends, then, on the space derivatives of $\mathbf{u}(\mathbf{r},t)$.

We introduce the newtonian law that the forces are proportional to accelerations. Thus, we assume that each component σ'_{ij} of $\boldsymbol{\sigma}'$ is a linear function of only the first derivatives of **u**, that is, of $\partial u_k/\partial x_l$ $(k, l = 1, 2, 3)$. Since $\boldsymbol{\sigma}'$ must vanish when **u** is a constant (no internal friction), there are no terms in **u**. If the velocity gradients are small, then higher-order derivatives and nonlinear terms need not be considered.

In order that the stress tensor be symmetric, the tangential components of $\boldsymbol{\sigma}'$ must contain derivatives of the form $\partial u_k/\partial x_l$, $k \neq l$, in symmetrical combinations, i.e., as

$$\frac{\partial u_k}{\partial x_l} + \frac{\partial u_l}{\partial x_k}$$

For an isotropic fluid, the most general form for the components of $\boldsymbol{\sigma}'$ satisfying the above conditions is

$$\sigma'_{ij} = \alpha\left(\frac{\partial u_i}{\partial x_j} + \frac{\partial u_j}{\partial x_i}\right) + \beta\left(\frac{\partial u_1}{\partial x_1} + \frac{\partial u_2}{\partial x_2} + \frac{\partial u_3}{\partial x_3}\right)\delta_{ij} \tag{6-31}$$

Since the fluid is isotropic, α and β are scalars, independent of velocity, and remain unchanged for each component of $\boldsymbol{\sigma}'$.

In dyadic form, Eq. (6-31) is

$$\boldsymbol{\sigma}' = 2\alpha \text{ sym } (\nabla \mathbf{u}) + \beta(\nabla \cdot \mathbf{u})\,\mathbf{1} \tag{6-32}$$

where sym (∇u) is the symmetrical part of the dyadic $\nabla\mathbf{u}$,

$$[\text{sym } (\nabla \mathbf{u})]_{ij} = \frac{1}{2}[(\nabla \mathbf{u})_{ij} + (\nabla \mathbf{u})_{ji}] = \frac{1}{2}\left(\frac{\partial u_j}{\partial x_i} + \frac{\partial u_i}{\partial x_j}\right) \tag{6-33}$$

It is convenient to write $\boldsymbol{\sigma}'$ as the sum of two parts,

$$\boldsymbol{\sigma}' = \mathbf{S} + \mathbf{T} \tag{6-34}$$

where

$$\mathbf{S} = \frac{1}{3}(\text{Tr } \boldsymbol{\sigma}')\,\mathbf{1} \tag{6-35}$$

and $\mathbf{T}$ is a dyadic with vanishing trace:

$$\text{Tr } \mathbf{T} = 0 \tag{6-36}$$

Taking $(1/3)(\sigma'_{11} + \sigma'_{22} + \sigma'_{33})$ from Eq. (6-31), we find that

$$\begin{aligned} S_{ii} &= (\tfrac{2}{3}\alpha + \beta)\nabla \cdot \mathbf{u} \\ S_{ij} &= 0 \qquad i \neq j \end{aligned} \tag{6-37}$$

Subtracting $\mathbf{S}$ from $\boldsymbol{\sigma}'$, we obtain $\mathbf{T}$,

$$T_{ij} = \alpha\left(\frac{\partial u_i}{\partial x_j} + \frac{\partial u_j}{\partial x_i}\right) - \frac{2}{3}(\nabla \cdot \mathbf{u})\,\delta_{ij} \tag{6-38}$$

or

$$\mathbf{T} = 2\alpha\left[\text{sym } (\nabla \mathbf{u}) - \frac{1}{3}(\nabla \cdot \mathbf{u})\,\mathbf{1}\right] \tag{6-39}$$

Replacing the coefficient α by η and $(\frac{2}{3}\alpha + \beta)$ by φ, we may write $\boldsymbol{\sigma}'$ as

$$\boldsymbol{\sigma}' = \varphi(\nabla \cdot \mathbf{u})\,\mathbf{1} + 2\eta\left[\text{sym } (\nabla \mathbf{u}) - \frac{1}{3}(\nabla \cdot \mathbf{u})\,\mathbf{1}\right] \tag{6-40}$$

and σ becomes

$$\boldsymbol{\sigma} = -\left[p + \left(\frac{2}{3}\eta - \varphi\right)\nabla \cdot \mathbf{u}\right]\mathbf{1} + 2\eta \operatorname{sym}(\nabla \mathbf{u}) \tag{6-41}$$

The constant η is the *coefficient of shear viscosity*, and φ is the *coefficient of bulk viscosity*. For an incompressible fluid, $\nabla \cdot \mathbf{u}$ vanishes and σ depends only on the coefficient of shear viscosity:

$$\boldsymbol{\sigma} = -p\mathbf{1} + 2\eta \operatorname{sym}(\nabla \mathbf{u}) \tag{6-42}$$

Since many liquids may be regarded as incompressible, η is generally regarded as the more important viscosity coefficient. For a general fluid not at equilibrium, the pressure P exerted by the fluid is given by

$$\begin{aligned} P &= -\frac{1}{3}\operatorname{Tr}\boldsymbol{\sigma} \\ &= p - \varphi\nabla \cdot \mathbf{u} \end{aligned} \tag{6-43}$$

Thus, φ represents the contribution of internal friction to the normal stress. Both η and φ are functions of pressure and temperature.

6-5 Equation of motion

The conservation of linear momentum in a fluid system leads to the equation of motion. If $\mathcal{G}$ is the x component of the total momentum of the system, then G is the x component of momentum per unit mass:

$$\begin{aligned} \mathcal{G} &= mu_x \\ G &= \frac{\mathcal{G}}{m} = u_x \end{aligned} \tag{6-44}$$

Since momentum is carried only by the actual flow of fluid, the x component of the momentum flux, according to Eq. (6-12), is $\rho u_x \mathbf{u}$.

Momentum is generated within the system because of volume forces of external fields and because of surface forces. If $\mathbf{X}$ is the external force per unit mass acting on the fluid system, the internal source of the x component of momentum due to volume forces is ρX_x. The surface force acting on the volume V is $\int_S \boldsymbol{\sigma} \cdot d\mathbf{s}$. This surface integral may be converted into a volume integral by the use of Gauss' theorem. The x component is, then,

$$\int_S \sum_{i=1}^{3} \sigma_{1i}\, ds_i = \int_V \sum_{i=1}^{3} \frac{\partial \sigma_{i1}}{\partial x_i}\, dv = \int_V (\nabla \cdot \sigma)_x\, dv \tag{6-45}$$

where the symmetrical character of the stress tensor has been used. The surface force contribution to ϕ_G is, then, $(\nabla \cdot \sigma)_x$, and the complete internal source of the x component of momentum is

$$\phi_G = \rho X_x + (\nabla \cdot \sigma)_x \tag{6-46}$$

Substituting these various quantities into the general conservation equation (6-16), we have

$$\frac{\partial}{\partial t}(\rho u_x) + \nabla \cdot (\rho u_x \mathbf{u}) = \rho X_x + (\nabla \cdot \sigma)_x \tag{6-47}$$

The same procedure is repeated for the y and z components of the momentum of the system, and expressions similar to Eq. (6-47) result. When the three equations for the three components of the total momentum are combined in a single vector equation, the equation of motion is obtained:

$$\frac{\partial}{\partial t}(\rho \mathbf{u}) + \nabla \cdot (\rho \mathbf{u}\mathbf{u}) = \rho \mathbf{X} + \nabla \cdot \sigma \tag{6-48}$$

To obtain the equation of motion in terms of the substantial derivative, we let G be u_x in Eq. (6-24) and obtain

$$\rho \frac{du_x}{dt} = \rho X_x + (\nabla \cdot \sigma)_x \tag{6-49}$$

Combining Eq. (6-49) with the corresponding expressions for the y and z components of the momentum into a single vector equation, we have

$$\rho \frac{d\mathbf{u}}{dt} = \rho \mathbf{X} + \nabla \cdot \sigma \tag{6-50}$$

We now specialize the equation of motion (6-50) to a newtonian fluid, i.e., to a fluid whose viscous forces are given by the newtonian stress tensor (6-41). The component σ_{ji} of σ is

$$\sigma_{ji} = -[p + (\tfrac{2}{3}\eta - \varphi)\nabla \cdot \mathbf{u}]\,\delta_{ji} + \eta\left(\frac{\partial u_i}{\partial x_j} + \frac{\partial u_j}{\partial x_i}\right) \tag{6-51}$$

Since the viscosity coefficients η and φ are independent of x, y, and z, the ith component of the divergence of the stress tensor $\boldsymbol{\sigma}$ is

$$\begin{aligned}(\nabla \cdot \sigma)_i &= \sum_j \frac{\partial \sigma_{ji}}{\partial x_j} \\ &= -\frac{\partial}{\partial x_i}[p + (\tfrac{2}{3}\eta - \varphi)\nabla \cdot \mathbf{u}] + \eta\nabla^2 u_i + \eta\frac{\partial}{\partial x_i}(\nabla \cdot \mathbf{u})\end{aligned} \tag{6-52}$$

where the relations

$$\sum_{j=1}^{3} \frac{\partial^2 u_i}{\partial x_j^2} = \nabla^2 u_i$$
$$\sum_{j=1}^{3} \frac{\partial u_j}{\partial x_j} = \nabla \cdot \mathbf{u} \tag{6-53}$$

have been used. The divergence of $\boldsymbol{\sigma}$ is then

$$\nabla \cdot \sigma = -\nabla p + (\frac{1}{3}\eta + \varphi)\nabla(\nabla \cdot \mathbf{u}) + \eta\nabla^2\mathbf{u} \tag{6-54}$$

Substitution of Eq. (6-54) into the equation of motion (6-50) yields

$$\rho\frac{d\mathbf{u}}{dt} = \rho\mathbf{X} - \nabla p + \eta\nabla^2\mathbf{u} + (\frac{1}{3}\eta + \varphi)\nabla(\nabla \cdot \mathbf{u}) \tag{6-55}$$

Equation (6-55) is called the *Navier-Stokes equation of motion.*

If the fluid is incompressible, we have $\nabla \cdot \mathbf{u} = 0$, and the Navier-Stokes equation becomes

$$\rho\frac{d\mathbf{u}}{dt} = \rho\mathbf{X} - \nabla p + \eta\nabla^2\mathbf{u} \tag{6-56}$$

For an ideal fluid, both η and φ vanish, and the equation of motion becomes

$$\rho\frac{d\mathbf{u}}{dt} = \rho\mathbf{X} - \nabla p \tag{6-57}$$

Equation (6-57) is known as *Euler's equation.*

6-6 Mechanical equilibrium and hydrostatics

A fluid is said to be in *mechanical equilibrium* if the velocity $\mathbf{u}$ of the local center of mass is independent of position $\mathbf{r}$ and of time t. For such a fluid, the Navier-Stokes equation of motion (6-55) becomes

$$\rho\mathbf{X} - \nabla p = 0 \tag{6-58}$$

The external forces are, therefore, balanced by the gradient of the pressure. When there are no external forces acting on a system which is in mechanical equilibrium, we have $\nabla p = 0$ or $p =$ constant, and the pressure of the fluid is uniform.

For a fluid in a gravitational field, the external force $\mathbf{X}$ per unit mass is a constant vector g, whose magnitude g is the acceleration due to gravity and whose direction is downward. When the fluid is also in mechanical equilibrium, Eq. (6-58) applies and takes the form

$$\nabla p = \rho \mathbf{g} \tag{6-59}$$

If we take the positive z axis as pointing directly upward, then Eq. (6-59) in component form is

$$\begin{aligned} \frac{\partial p}{\partial x} &= \frac{\partial p}{\partial y} = 0 \\ \frac{\partial p}{\partial z} &= -\rho g \end{aligned} \tag{6-60}$$

Thus, the pressure varies only with the height z. When solved for ρ, Eq. (6-60) becomes

$$\rho = -\frac{1}{g}\frac{\partial p}{\partial z} \tag{6-61}$$

and we see that ρ is a function of z only.

If the density of the fluid is essentially uniform throughout its volume, then ρ in Eq. (6-60) may be regarded as a constant and the equation integrated to yield

$$p = -\rho g z + \text{const} \tag{6-62}$$

If p has the value p_0 when z equals z_0, Eq. (6-62) becomes

$$p = p_0 - \rho g (z - z_0) \tag{6-63}$$

For a fluid system in which the density varies with height, we introduce the thermodynamic relation

$$dF = -S\,dT + \frac{dp}{\rho} \tag{6-64}$$

where F is the specific Gibbs free energy and S is the specific entropy. If the temperature of the system is uniform, then dT is zero and

$$\frac{dp}{\rho} = dF \tag{6-65}$$

Substituting this expression into Eq. (6-61), we have

$$d(F + gz) = 0 \tag{6-66}$$

or

$$F + gz = \text{const} \tag{6-67}$$

Equation (6-67) is the condition for thermodynamic equilibrium in a gravitational field.

6-7 Energy transport equation

The conservation of energy in a fluid system leads to the energy transport equation. The total energy of a volume V of fluid relative to an arbitrary state is the sum of the kinetic energy $\frac{1}{2}\rho V u^2$ and the internal energy. If E is the specific internal energy (internal energy per unit mass) for the fluid, then G in Eq. (6-16) is

$$G = E + \tfrac{1}{2}u^2 \tag{6-68}$$

The current density or flux of energy consists of two parts, the convection current and the conduction current. The convection current is that part of the energy flux associated with the motion of the local center of mass. According to Eq. (6-12), this contribution to the energy flux is $\rho(E + \frac{1}{2}u^2)\mathbf{u}$. The conduction current density includes energy flow due to diffusion and to heat flow and is given the symbol $\mathbf{j}_E$. Combining these two contributions to $\mathbf{J}_G$, we have

$$\mathbf{J}_G = \rho(E + \tfrac{1}{2}u^2)\mathbf{u} + \mathbf{j}_E \tag{6-69}$$

The internal production ϕ_G of energy per unit volume and time is equal to the rate of work done by volume and surface forces on a fluid element of unit volume. The rate at which work is done on the element by the volume force due to external force fields is just the product of the force and the distance traveled per unit time,

$$\sum_{\alpha=1}^{\nu} \rho_\alpha \mathbf{X}_\alpha \cdot \mathbf{u}_\alpha$$

where $\mathbf{X}_\alpha$ is the external force acting on component α per unit mass of component α. The total volume force per unit volume is the sum of the contributions from each component,

$$\rho\mathbf{X} = \sum_{\alpha=1}^{\nu} \rho_\alpha \mathbf{X}_\alpha \tag{6-70}$$

If the diffusion fluxes $\mathbf{j}_\alpha$ are introduced through Eq. (6-6), then the rate of work production from volume forces becomes

$$\sum_{\alpha=1}^{\nu} \mathbf{j}_\alpha \cdot \mathbf{X}_\alpha + \rho\mathbf{u} \cdot \mathbf{X}$$

The rate at which surface forces perform work on the surface S of a volume V is the product of the surface force and the velocity of the local center of mass integrated over each element of surface:

$$\oint_S (\mathbf{u} \cdot \boldsymbol{\sigma}) \cdot d\mathbf{s}$$

When Gauss' theorem is applied, this expression becomes

$$\int_V \nabla \cdot (\mathbf{u} \cdot \boldsymbol{\sigma})\, dv$$

Thus, the contribution to ϕ_G of surface forces is $\nabla \cdot (\mathbf{u} \cdot \boldsymbol{\sigma})$ and ϕ_G takes the form

$$\phi_G = \sum_{\alpha=1}^{\nu} \mathbf{j}_\alpha \cdot \mathbf{X}_\alpha + \rho \mathbf{u} \cdot \mathbf{X} + \nabla \cdot (\mathbf{u} \cdot \boldsymbol{\sigma}) \tag{6-71}$$

Substitution of Eqs. (6-68), (6-69), and (6-71) into the general conservation equation (6-16) gives the energy transport equation

$$\frac{\partial}{\partial t} [\rho(E + \tfrac{1}{2} u^2)] + \nabla \cdot [\rho(E + \tfrac{1}{2} u^2)\mathbf{u}]$$

$$= \sum_{\alpha=1}^{\nu} \mathbf{j}_\alpha \cdot \mathbf{X}_\alpha + \rho \mathbf{u} \cdot \mathbf{X} + \nabla \cdot (\mathbf{u} \cdot \boldsymbol{\sigma}) - \nabla \cdot \mathbf{j}_E \tag{6-72}$$

To obtain the energy transport equation in terms of the substantial derivative, we substitute G in Eq. (6-68), $\mathbf{J}_G$ in Eq. (6-69), and ϕ_G in Eq. (6-71) into the general conservation equation (6-24) and obtain

$$\rho \frac{d}{dt}(E + \tfrac{1}{2} u^2) + \nabla \cdot \mathbf{j}_E = \sum_{\alpha=1}^{\nu} \mathbf{j}_\alpha \cdot \mathbf{X}_\alpha + \rho \mathbf{u} \cdot \mathbf{X} + \nabla \cdot (\mathbf{u} \cdot \boldsymbol{\sigma}) \tag{6-73}$$

The quantity $\nabla \cdot (\mathbf{u} \cdot \boldsymbol{\sigma})$ may be modified in the following way:

$$\begin{aligned}
\nabla \cdot (\mathbf{u} \cdot \boldsymbol{\sigma}) &= \sum_{j=1}^{3} \frac{\partial}{\partial x_j} \sum_{i=1}^{3} u_i \sigma_{ij} \\
&= \sum_{i=1}^{3} \sum_{j=1}^{3} \frac{\partial}{\partial x_j}(u_i \sigma_{ij}) \\
&= \sum_{i=1}^{3} \sum_{j=1}^{3} u_i \frac{\partial \sigma_{ji}}{\partial x_j} + \sum_{i=1}^{3} \sum_{j=1}^{3} \sigma_{ij} \frac{\partial u_i}{\partial x_j} \\
&= \mathbf{u} \cdot (\nabla \cdot \boldsymbol{\sigma}) + \boldsymbol{\sigma} : \nabla \mathbf{u}
\end{aligned} \tag{6-74}$$

where the symmetry property of the stress tensor $\boldsymbol{\sigma}$ is used. Substitution of Eq. (6-74) and the relation

$$\rho \frac{d}{dt}(\tfrac{1}{2} u^2) = \rho \mathbf{u} \cdot \frac{d\mathbf{u}}{dt}$$

into Eq. (6-73) gives

$$\rho \frac{dE}{dt} - \sigma\text{:}\nabla \mathbf{u} - \sum_{\alpha=1}^{\nu} \mathbf{j}_\alpha \cdot \mathbf{X}_\alpha + \nabla \cdot \mathbf{j}_E = \mathbf{u} \cdot \left[\rho \mathbf{X} + \nabla \cdot \sigma - \rho \frac{d\mathbf{u}}{dt}\right]$$

the right-hand side of which vanishes by the equation of motion (6-50), giving the desired result:

$$\rho \frac{dE}{dt} = \sigma\text{:}\nabla \mathbf{u} + \sum_{\alpha=1}^{\nu} \mathbf{j}_\alpha \cdot \mathbf{X}_\alpha - \nabla \cdot \mathbf{j}_E \tag{6-75}$$

6-8 Entropy transport equation and the second law of thermodynamics

In this section we derive a transport equation for the entropy of a fluid system. If S is the specific entropy, the general conservation equation (6-24) takes the form

$$\rho \frac{dS}{dt} + \nabla \cdot (\mathbf{J}_S - \rho S \mathbf{u}) = \phi_S \tag{6-76}$$

The quantity $\mathbf{J}_S - \rho S\mathbf{u}$ is the current density of the entropy due only to diffusion and heat flow. The convective current density of entropy, $\rho S\mathbf{u}$, is subtracted. If we give the symbol $\mathbf{j}_S$ to the entropy flux due to diffusion and heat flow, Eq. (6-76) becomes

$$\rho \frac{dS}{dt} = \phi_S - \nabla \cdot \mathbf{j}_S \tag{6-77}$$

We now introduce the assumption of local equilibrium:
For a nonequilibrium system all thermodynamic functions of state exist for each element of the system and these thermodynamic quantities are the same functions of the local state variables as the corresponding equilibrium thermodynamic quanties. Thus we may consider the temperature $T(\mathbf{r},t)$, the pressure $p(\mathbf{r},t)$, the specific entropy $S(\mathbf{r},t)$, the chemical potential $\mu_\alpha(\mathbf{r},t)$ of component α, etc., for the fluid. Furthermore, these quantities are related by the same equations as those for systems in equilibrium. In particular, we may consider the Gibbs equation for an open system:

$$dE = T\,dS - p\,d(1/\rho) + \sum_{\alpha=1}^{\nu} \mu_\alpha\,dx_\alpha \tag{6-78}$$

This equation is written in terms of specific quantities, i.e., quantities per unit mass. Note that $1/\rho$ is the volume per unit mass.

The rate of entropy production may be obtained by dividing Eq. (6-78) by *dt* and rearranging to give

$$\rho \frac{dS}{dt} = \frac{\rho}{T}\frac{dE}{dt} - \frac{p}{\rho T}\frac{d\rho}{dt} - \frac{\rho}{T}\sum_{\alpha=1}^{\nu} \mu_\alpha \frac{dx_\alpha}{dt} \tag{6-79}$$

The quantities dE/dt, $d\rho/dt$, and dx_α/dt are given by Eqs. (6-75), (6-26), and (6-25), respectively. The problem is to substitute these three expressions into Eq. (6-79) and then to rearrange the result into the form of Eq. (6-77).

The substitution and rearrangement are straightforward. The result is that

$$T\phi_S = (\boldsymbol{\sigma} + p\mathbf{1}) : \nabla\mathbf{u} - \mathbf{j}_E \cdot \nabla \ln T - \sum_{\alpha=1}^{\nu} \mathbf{j}_\alpha \cdot [T\nabla(\mu_\alpha/T) - \mathbf{X}_\alpha] + T\phi^{(\text{react})} \tag{6-80}$$

$$\mathbf{j}_S = \frac{\mathbf{j}_E - \sum_{\alpha=1}^{\nu} \mu_\alpha \mathbf{j}_\alpha}{T} \tag{6-81}$$

where we have defined

$$\phi^{(\text{react})} = - \sum_{\alpha=1}^{\nu} \mu_\alpha \phi_\alpha \tag{6-82}$$

and have used the identities

$$\nabla \cdot \mathbf{j}_E = T\nabla \cdot (\mathbf{j}_E/T) + \mathbf{j}_E \cdot \nabla \ln T \tag{6-83}$$

$$\frac{\mu_\alpha}{T} \nabla \cdot \mathbf{j}_\alpha = \nabla \cdot \left(\frac{\mu_\alpha \mathbf{j}_\alpha}{T}\right) - \mathbf{j}_\alpha \cdot \nabla\left(\frac{\mu_\alpha}{T}\right) \tag{6-84}$$

$$\nabla \cdot \mathbf{u} = \mathbf{1} : \nabla\mathbf{u} \tag{6-85}$$

For an irreversible process in a *closed system*, the second law of thermodynamics states that the change in entropy of the system must be greater than the heat absorbed by the system during the irreversible change divided by the absolute temperature T_{surr} of the surroundings,

$$dS > \frac{dQ}{T_{\text{surr}}} \tag{6-86}$$

where dQ is the differential heat absorbed per unit mass of the system. We now write the entropy change in the form

$$\begin{aligned} dS &= d_eS + d_iS \\ d_eS &= \frac{dQ}{T_{\text{surr}}} \end{aligned} \tag{6-87}$$

where d_eS is the change in entropy due to heat absorbed from the surroundings and d_iS is the entropy production within the system. We can see immediately from Eqs. (6-86) and (6-87) that the second law states

$$d_iS > 0 \tag{6-88}$$

In an open system, described by Eqs. (6-77), (6-80), and (6-81), the quantity d_eS is no longer just dQ/T_{surr}. A term must be added to account for the entropy change resulting when matter is transported into and out of the system, as shown in Eq. (6-81). It is then possible for dS and d_eS to be either positive or negative and, in general, Eq. (6-86) does not hold for open systems. The proper statement of the second law of thermodynamics for open systems is that the internal entropy production d_iS or its rate ϕ_S due to irreversible processes in the system is positive:

$$\phi_S > 0 \tag{6-89}$$

6-9 Sound waves

In this section we consider small oscillations in the local density and pressure for an ideal, compressible fluid. The reference state is a uniform fluid at equilibrium with pressure p_0 and density ρ_0. The local pressure $p(\mathbf{r},t)$ and density $\rho(\mathbf{r},t)$ oscillate about their reference-state values as the result of a disturbance. Thus, at each point the fluid experiences an alternating compression and rarefaction. Such oscillatory motion in a compressible fluid is called a *sound wave.*

The local pressure and density may be written

$$\begin{aligned} p &= p_0 + \tilde{p} \\ \rho &= \rho_0 + \tilde{\rho} \end{aligned} \tag{6-90}$$

where $\tilde{p}, \tilde{\rho}$ are the deviations from the reference-state values. Both $\tilde{p}, \tilde{\rho}$ are very small compared with p_0, ρ_0, so that $\tilde{p} \ll p_0$, $\tilde{\rho} \ll \rho_0$. As an example, for a sound wave of moderate intensity in air, $\tilde{p}$ is on the order of 10^{-7} atm with p_0 equal to 1 atm.

If the fluid is at uniform temperature and composition, there are no heat flows or diffusion currents. The deviations in pressure are considered sufficiently small that they do not lead to gradients in the chemical potentials. Therefore, ϕ_S

and $\mathbf{j}_S$ in Eqs. (6-80) and (6-81) are zero and the fluid is *isentropic* (at constant entropy).

The pressure p may be regarded as a function of the density ρ. If we expand $p(\rho)$ as a Taylor series about ρ_0 and keep only the linear term in $\tilde{\rho}$ (neglecting the higher-order terms), we have

$$p = p(\rho_0) + \left(\frac{\partial p}{\partial \rho}\right)_S \tilde{\rho} \tag{6-91}$$

Since $p(\rho_0)$ is p_0, Eqs. (6-90) and (6-91) give

$$\tilde{p} = \left(\frac{\partial p}{\partial \rho}\right)_S \tilde{\rho} \tag{6-92}$$

The velocity $\mathbf{u}(\mathbf{r},t)$ of the local center of mass for the fluid system under consideration here is also a small quantity. Thus, any second-order term which depends on the product of $\tilde{\rho}$ and $\mathbf{u}$, of $\tilde{p}$ and $\mathbf{u}$, or of $\mathbf{u}$ and $\mathbf{u}$ in a mathematical equation may be neglected. Recognizing that ρ_0 is a constant and neglecting a second-order term, the equation of continuity (6-20) becomes

$$\frac{\partial \tilde{\rho}}{\partial t} + \rho_0 \nabla \cdot \mathbf{u} = 0 \tag{6-93}$$

We may eliminate $\tilde{\rho}$ in favor of $\tilde{p}$ through Eq. (6-92) and obtain

$$\frac{\partial \tilde{p}}{\partial t} + \rho_0 \left(\frac{\partial p}{\partial \rho}\right)_S \nabla \cdot \mathbf{u} = 0 \tag{6-94}$$

Euler's equation (6-57) is the appropriate form here for the equation of motion. Since there are no external forces on the system and second-order terms are negligible, this equation becomes

$$\rho_0 \frac{\partial \mathbf{u}}{\partial t} + \nabla \tilde{p} = 0 \tag{6-95}$$

We now differentiate Eq. (6-94) with respect to time,

$$\frac{\partial^2 \tilde{p}}{\partial t^2} + \rho_0 \left(\frac{\partial p}{\partial \rho}\right)_S \nabla \cdot \frac{\partial \mathbf{u}}{\partial t} = 0 \tag{6-96}$$

and use Eq. (6-95) to eliminate $\partial \mathbf{u}/\partial t$. The result is

$$\nabla^2 \tilde{p} - \frac{1}{(\partial p/\partial \rho)_S} \frac{\partial^2 \tilde{p}}{\partial t^2} = 0 \tag{6-97}$$

Thus $\tilde{p}$ obeys the wave equation (4-65) with the velocity c of the sound wave given by

$$c = \sqrt{(\partial p/\partial \rho)_S} \tag{6-98}$$

Since $\tilde{\rho}$ is directly proportional to $\tilde{p}$, $\tilde{\rho}$ also obeys the wave equation with the same value for the velocity.

We next consider a sound wave for which $\tilde{p}$ and $\tilde{\rho}$ depend only on one coordinate, say z; hence, $\tilde{p}(z,t)$ and $\tilde{\rho}(z,t)$. From the discussion in Sec. 4-9, it is apparent that $\tilde{p}(z,t)$ and $\tilde{\rho}(z,t)$ are plane waves. The behavior of $\mathbf{u}(\mathbf{r},t)$ may be obtained from Eq. (6-95), which reduces to

$$\rho_0 \frac{\partial \mathbf{u}}{\partial t} = -\mathbf{k}\frac{\partial \tilde{p}}{\partial z} \tag{6-99}$$

Since $\mathbf{u}$ is zero in the reference state, Eq. (6-99) shows that $\mathbf{u}$ is in the z direction. Thus, the velocity of the local center of mass in a sound wave is in the direction of propagation. Such a vector is said to be *longitudinal*. This behavior is in contrast to the transverse nature of the electric field vector $\mathbf{E}$ and the magnetic field vector $\mathbf{H}$ for a plane wave as discussed in Sec. 4-9.

The velocity c of the sound wave takes a particularly simple form for an ideal gas. Since the fluid system remains isentropic, the propagation of the sound wave is an adiabatic process and the relation

$$pV^{\gamma} = \text{const} \tag{6-100}$$

applies, where γ is the ratio of the heat capacity at constant pressure to the heat capacity at constant volume:

$$\gamma = \frac{C_P}{C_V} \tag{6-101}$$

Since $\rho = m/V$, where m is the mass of the system, we have

$$p = \text{const}\,\rho^{\gamma} \tag{6-102}$$

$$\left(\frac{\partial p}{\partial \rho}\right)_S = \frac{\gamma p}{\rho} \tag{6-103}$$

From Eq. (6-98) we then obtain

$$c = \sqrt{\gamma p/\rho} \tag{6-104}$$

The equation of state $pV = nRT$ for the ideal gas may be written in the form

$$p = \frac{\rho RT}{M} \tag{6-105}$$

where R is the universal gas constant and M is the molecular weight of the gas. The velocity of sound then takes the form

$$c = \sqrt{\gamma RT/M} \tag{6-106}$$

PROBLEMS

1. By setting the total force and the total torque equal to zero, show that the stress tensor is symmetric for an element of fluid which is in static equilibrium or is moving with a constant velocity.

2. Show that the Navier-Stokes equation of motion may be expressed in the form

$$\rho \frac{d\mathbf{u}}{dt} = \rho\mathbf{X} - \nabla p - \eta\nabla \times (\nabla \times \mathbf{u}) + (\frac{4}{3}\eta + \varphi)\nabla(\nabla \cdot \mathbf{u})$$

3. Derive the flow equation

$$\frac{\partial \mathbf{\Omega}}{\partial t} + \nabla \times (\mathbf{\Omega} \times \mathbf{u}) = \frac{\eta}{\rho} \nabla^2 \mathbf{\Omega}$$

for an incompressible fluid with conservative external forces, where the *vorticity* $\mathbf{\Omega}$ is defined as $\mathbf{\Omega} = \nabla \times \mathbf{u}$. [*Hint*: Use Eqs. (3-44) and (3-48) and the equation of motion.]

4. By setting the conduction current density of heat $\mathbf{j}_E$ proportional to the temperature gradient ($\mathbf{j}_E = -\kappa\nabla T$) and using the thermodynamic relation

$$dE = C_V\, dT + \left[T\left(\frac{\partial p}{\partial T}\right)_V - p\right] d(1/\rho)$$

where C_V is the heat capacity per unit mass at constant volume, derive the following differential equation for the temperature distribution in the steady state (time-independent, nonequilibrium state) for an ideal gas with no external forces:

$$\rho C_V \mathbf{u} \cdot \nabla T = \kappa\nabla^2 T - p(\nabla \cdot \mathbf{u})$$

5. (*a*) Show that in terms of the density ρ the van der Waals equation of state is

$$\left(p + \frac{a\rho^2}{M^2}\right)\left(\frac{M}{\rho} - b\right) = RT$$

where a and b are the usual van der Waals constants and M is the molecular weight.

(*b*) Using the thermodynamic relation

$$\left(\frac{\partial p}{\partial \rho}\right)_S = \gamma\left(\frac{\partial p}{\partial \rho}\right)_T$$

find the velocity of sound for a van der Waals gas.

(*c*) Show that the expression derived in part (*b*) reduces to Eq. (6-106) when *a* and *b* vanish.

Chapter seven

Vectors in Spherical Coordinates

7-1 Spherical coordinate system

Up to this point we have considered vectors represented only in a cartesian coordinate system. However, the representation of vectors and vector analysis are not limited to rectangular coordinates. In certain applications of vectors it is often convenient to use a spherical or polar coordinate system, in which the location of a point is given by its position on the surface of a sphere with a particular radius. Such a spherical coordinate system is illustrated in Fig. 7-1. The vector **A** is determined by three coordinates: the radius r of the sphere, the angle θ which the vector **A** makes with the positive z axis, and the angle ϕ which the projection of the vector on the xy plane makes with the x axis. Since $r \cos\theta$ is the projection of **A** on the z axis and $r \sin\theta$ the projection of **A** on the xy plane, the cartesian coordinates x, y, z and the spherical coordinates r, θ, ϕ are related by

$$
\begin{aligned}
x &= r \sin\theta \cos\phi \\
y &= r \sin\theta \cos\phi \\
z &= r \cos\theta \\
r &= (x^2 + y^2 + z^2)^{1/2} \\
\theta &= \cos^{-1} \frac{z}{r} \\
\phi &= \tan^{-1} \frac{y}{x}
\end{aligned}
\qquad (7\text{-}1)
$$

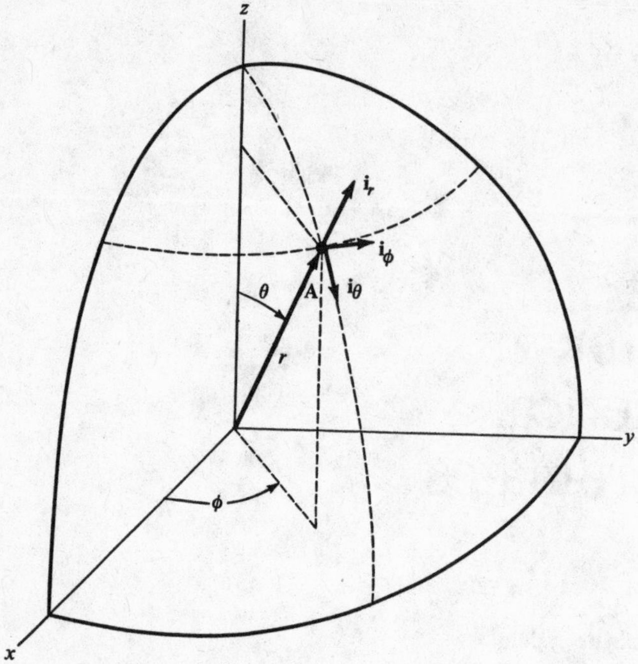

FIGURE 7-1

We next introduce a set of unit vectors $\mathbf{i}_r, \mathbf{i}_\theta, \mathbf{i}_\phi$ for the spherical coordinates r, θ, ϕ. The unit vector $\mathbf{i}_r$ corresponding to the radial coordinate r is directed along the positive r direction, i.e., along the radius of the sphere and directed outward. The unit vector $\mathbf{i}_\theta$ corresponding to the angle θ is tangent to the sphere of radius r and is pointed in the direction of increasing θ. Similarly, the unit vector $\mathbf{i}_\phi$ is tangent to the sphere and is pointed in the direction of increasing ϕ. All three unit vectors as defined are shown in Fig. 7-1.

By virtue of these definitions, the three unit vectors $\mathbf{i}_r, \mathbf{i}_\theta, \mathbf{i}_\phi$ are mutually orthogonal (perpendicular). With the angles defined as in Fig. 7-1 with respect to a right-handed cartesian coordinate system, the ordered set $(\mathbf{i}_r, \mathbf{i}_\theta, \mathbf{i}_\phi)$ form a right-handed system, so that

$$\begin{aligned} \mathbf{i}_r \times \mathbf{i}_\theta &= \mathbf{i}_\phi \\ \mathbf{i}_\theta \times \mathbf{i}_\phi &= \mathbf{i}_r \\ \mathbf{i}_\phi \times \mathbf{i}_r &= \mathbf{i}_\theta \end{aligned} \tag{7-2}$$

The unit vectors $\mathbf{i}_r, \mathbf{i}_\theta, \mathbf{i}_\phi$ all change directions from point to point on a given sphere. Thus, at a point P_1 each of these unit vectors has a particular orientation with respect to the fixed cartesian coordinate system. At some other point P_2 on the same sphere of radius r, each of the unit vectors has a different orientation in the xyz frame of reference. This behavior is in contrast with the cartesian unit vectors $\mathbf{i}, \mathbf{j}, \mathbf{k}$. The unit vector $\mathbf{i}$ is always along the direction of the positive x axis regardless of the point under consideration. Similarly, the unit vectors $\mathbf{j}$ and $\mathbf{k}$ are always along the positive y and z axes, respectively. Note, however, that at any point the unit vectors $\mathbf{i}_r, \mathbf{i}_\theta, \mathbf{i}_\phi$ are mutually orthogonal. This requirement is essential for the statements of general vector relationships.

A vector $\mathbf{A}$ may be expressed in terms of its components in the spherical coordinate system,

$$\mathbf{A} = \mathbf{i}_r A_r + \mathbf{i}_\theta A_\theta + \mathbf{i}_\phi A_\phi \tag{7-3}$$

where A_r, A_θ, A_ϕ are the r, θ, ϕ components, respectively. The relationships between the cartesian components A_x, A_y, A_z and the spherical components A_r, A_θ, A_ϕ may be readily derived. First, the relationships involving the unit vectors are obtained. Since the projection of $\mathbf{i}_r$ on the positive z axis is $\cos\theta$ and on the xy plane is $\sin\theta$, we have

$$\mathbf{i}_r = \mathbf{i}\sin\theta\cos\phi + \mathbf{j}\sin\theta\sin\phi + \mathbf{k}\cos\theta \tag{7-4a}$$

Since the angle between $\mathbf{i}_\theta$ and the positive z axis is $(\pi/2) + \theta$, we have

$$\begin{aligned}\mathbf{i}_\theta &= \mathbf{i}\sin\left(\frac{\pi}{2}+\theta\right)\cos\phi + \mathbf{j}\sin\left(\frac{\pi}{2}+\theta\right)\sin\phi + \mathbf{k}\cos\left(\frac{\pi}{2}+\theta\right)\\ &= \mathbf{i}\cos\theta\cos\phi + \mathbf{j}\cos\theta\sin\phi - \mathbf{k}\sin\theta\end{aligned} \tag{7-4b}$$

Since $\mathbf{i}_\phi$ is parallel to the xy plane and therefore has no z component, we have

$$\mathbf{i}_\phi = -\mathbf{i}\sin\phi + \mathbf{j}\cos\phi \tag{7-4c}$$

The unit vectors $\mathbf{i}_r, \mathbf{i}_\theta, \mathbf{i}_\phi$ in Eqs. (7-4) satisfy Eqs. (7-2), which can also be used to determine one of the unit vectors when the other two are known. If we now substitute Eqs. (7-4) into Eq. (7-3), we obtain

$$\begin{aligned}\mathbf{A} = {} & \mathbf{i}(\sin\theta\cos\phi\, A_r + \cos\theta\cos\phi\, A_\theta - \sin\phi\, A_\phi)\\ & + \mathbf{j}(\sin\theta\sin\phi\, A_r + \cos\theta\sin\phi\, A_\theta + \cos\phi\, A_\phi)\\ & + \mathbf{k}(\cos\theta\, A_r - \sin\theta\, A_\theta)\end{aligned} \tag{7-5}$$

Therefore, A_x, A_y, A_z are related to A_r, A_θ, A_ϕ by

$$\begin{aligned}A_x &= \sin\theta\cos\phi\, A_r + \cos\theta\cos\phi\, A_\theta - \sin\phi\, A_\phi\\ A_y &= \sin\theta\sin\phi\, A_r + \cos\theta\sin\phi\, A_\theta + \cos\phi\, A_\phi\\ A_z &= \cos\theta\, A_r - \sin\theta\, A_\theta\end{aligned} \tag{7-6}$$

Equations (7-4) may be solved for **i**, **j**, **k** to yield

$$\begin{aligned}\mathbf{i} &= \sin\theta\ \cos\phi\ \mathbf{i}_r + \cos\theta\ \cos\phi\ \mathbf{i}_\theta - \sin\phi\ \mathbf{i}_\phi \\ \mathbf{j} &= \sin\theta\ \sin\phi\ \mathbf{i}_r + \cos\theta\ \sin\phi\ \mathbf{i}_\theta + \cos\phi\ \mathbf{i}_\phi \\ \mathbf{k} &= \cos\theta\ \mathbf{i}_r - \sin\theta\ \mathbf{i}_\theta\end{aligned} \tag{7-7}$$

Similarly, Eqs. (7-6) may be inverted:

$$\begin{aligned}A_r &= \sin\theta\ \cos\phi\ A_x + \sin\theta\ \sin\phi\ A_y + \cos\theta\ A_z \\ A_\theta &= \cos\theta\ \cos\phi\ A_x + \cos\theta\ \sin\phi\ A_y - \sin\theta\ A_z \\ A_\phi &= -\sin\phi\ A_x + \cos\phi\ A_y\end{aligned} \tag{7-8}$$

Note that the coefficients which relate $\mathbf{i}_r$, $\mathbf{i}_\theta$, $\mathbf{i}_\phi$ to **i**, **j**, **k** in Eqs. (7-4) are the same as those which relate A_r, A_θ, A_ϕ to A_x, A_y, A_z in Eqs. (7-8). Likewise, the coefficients in Eqs. (7-7) which relate **i**, **j**, **k** to $\mathbf{i}_r$, $\mathbf{i}_\theta$, $\mathbf{i}_\phi$ are identical to those in Eqs. (7-6) which relate A_x, A_y, A_z to A_r, A_θ, A_ϕ.

Since the unit vectors $\mathbf{i}_r$, $\mathbf{i}_\theta$, $\mathbf{i}_\phi$ change directions from point to point, it is not possible to add or subtract vectors when they are expressed in spherical coordinates. To illustrate this point, consider two arbitrary vectors **A** and **B** expressed in spherical coordinates:

$$\begin{aligned}\mathbf{A} &= \mathbf{i}_r A_r + \mathbf{i}_\theta A_\theta + \mathbf{i}_\phi A_\phi \\ \mathbf{B} &= \mathbf{i}_r' B_r + \mathbf{i}_\theta' B_\theta + \mathbf{i}_\phi' B_\phi\end{aligned} \tag{7-9}$$

Since, in general, **A** is not parallel to **B**, the unit vectors $\mathbf{i}_r$, $\mathbf{i}_\theta$, $\mathbf{i}_\phi$ are not parallel, respectively, to the unit vectors $\mathbf{i}_r'$, $\mathbf{i}_\theta'$, $\mathbf{i}_\phi'$. Thus, the sum of **A** and **B**,

$$\mathbf{A} + \mathbf{B} = (\mathbf{i}_r A_r + \mathbf{i}_r' B_r) + (\mathbf{i}_\theta A_\theta + \mathbf{i}_\theta' B_\theta) + (\mathbf{i}_\phi A_\phi + \mathbf{i}_\phi' B_\phi) \tag{7-10}$$

cannot be further simplified in terms of spherical coordinates. If Eqs. (7-4) and (7-8) are introduced to express $\mathbf{i}_r$, $\mathbf{i}_\theta$, $\mathbf{i}_\phi$, $\mathbf{i}_r'$, $\mathbf{i}_\theta'$, $\mathbf{i}_\phi'$, A_r, A_θ, A_ϕ, B_r, B_θ, B_ϕ in terms of **i**, **j**, **k**, A_x, A_y, A_z, B_x, B_y, B_z, Eq. (7-10) becomes

$$\mathbf{A} + \mathbf{B} = \mathbf{i}(A_x + B_x) + \mathbf{j}(A_y + B_y) + \mathbf{k}(A_z + B_z) \tag{1-4}$$

Similarly, the scalar product $\mathbf{A} \cdot \mathbf{B}$ and the vector product $\mathbf{A} \times \mathbf{B}$ cannot be conveniently formed in spherical coordinates.

7-2 Volume element

In cartesian coordinates the volume dv of a differential volume element with sides

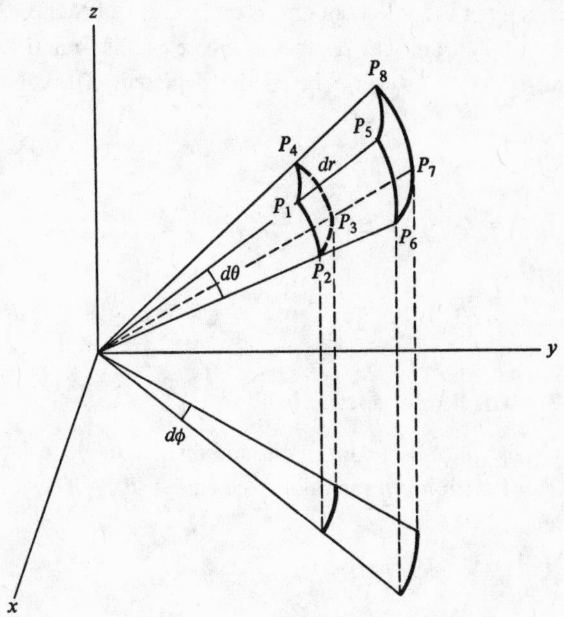

FIGURE 7-2

dx, dy, dz is simply $dv = dx\,dy\,dz$. In spherical coordinates the volume element dv is a slightly more elaborate expression. A volume element in spherical coordinates is illustrated in Fig. 7-2, where $P_1, \ldots, P_8$ indicate corners of the element. The various faces of the element are determined as follows:

$$\begin{aligned}
P_1P_2P_3P_4: &\qquad r = \text{const}\\
P_5P_6P_7P_8: &\qquad r + dr = \text{const}\\
P_1P_5P_8P_4: &\qquad \theta = \text{const}\\
P_2P_6P_7P_3: &\qquad \theta + d\theta = \text{const}\\
P_1P_2P_6P_5: &\qquad \phi = \text{const}\\
P_4P_3P_7P_8: &\qquad \phi + d\phi = \text{const}
\end{aligned}$$

In the limit $dr \to 0$, $d\theta \to 0$, $d\phi \to 0$, the curved edges of the element approach straight lines and the element itself approaches a rectangular parallelepiped. Thus, the volume dv is the product of the three edges (P_1P_5), (P_1P_2), (P_1P_4):

$$\begin{aligned}
dv &= dr \cdot r\,d\theta \cdot r \sin\theta\, d\phi\\
&= r^2 \sin\theta\, dr\, d\theta\, d\phi
\end{aligned} \tag{7-11}$$

The volume of a sphere may be obtained by integrating dv over the range of allowed values for r, θ, and ϕ. For a sphere of radius a, r ranges from 0 to a. It is customary to let θ vary from 0 to π. The angle ϕ must then cover the values from

0 to 2π in order that all points on a sphere of radius r be represented. (One could let θ vary from 0 to 2π, in which case ϕ runs from 0 to π. This choice leads to difficulties, however, when dv is integrated.) The volume V of the sphere is, then,

$$
\begin{aligned}
V &= \int_0^{2\pi} \int_0^{\pi} \int_0^{a} r^2 \sin\theta \, dr \, d\theta \, d\phi \\
&= 2\pi \cdot [-\cos\theta]_0^{\pi} \cdot \frac{a^3}{3} \\
&= \frac{4}{3}\pi a^3 \qquad (7\text{-}12)
\end{aligned}
$$

7-3 Gradient of a scalar field

Let $\psi(r,\theta,\phi)$ be a continuous and differentiable scalar function of position in space. This function can vary along the curves (P_1P_5), (P_1P_2), and (P_1P_4) in Fig. 7-2. If we let

$$
\begin{aligned}
(P_1P_5) &= dl_r &&= dr \\
(P_1P_2) &= dl_\theta &&= r\,d\theta \qquad (7\text{-}13) \\
(P_1P_4) &= dl_\phi &&= r\sin\theta \; d\phi
\end{aligned}
$$

the space derivatives of $\psi(r,\theta,\phi)$ in spherical coordinates are $\partial\psi/\partial l_r$, $\partial\psi/\partial l_\theta$, $\partial\psi/\partial l_\phi$. The gradient of $\psi(r,\theta,\phi)$ is, then, a vector in spherical coordinates with components $\partial\psi/\partial l_r$, $\partial\psi/\partial l_\theta$, $\partial\psi/\partial l_\phi$,

$$
\begin{aligned}
\nabla\psi &= \mathbf{i}_r \frac{\partial\psi}{\partial l_r} + \mathbf{i}_\theta \frac{\partial\psi}{\partial l_\theta} + \mathbf{i}_\phi \frac{\partial\psi}{\partial l_\phi} \\
&= \mathbf{i}_r \frac{\partial\psi}{\partial r} + \mathbf{i}_\theta \frac{1}{r}\frac{\partial\psi}{\partial\theta} + \mathbf{i}_\phi \frac{1}{r\sin\theta}\frac{\partial\psi}{\partial\phi} \qquad (7\text{-}14)
\end{aligned}
$$

In spherical coordinates the vector differential operator ∇ has the form, therefore,

$$
\nabla = \mathbf{i}_r \frac{\partial}{\partial r} + \mathbf{i}_\theta \frac{1}{r}\frac{\partial}{\partial\theta} + \mathbf{i}_\phi \frac{1}{r\sin\theta}\frac{\partial}{\partial\phi} \qquad (7\text{-}15)
$$

7-4 Divergence of a vector field

In calculating the divergence of a vector field $\mathbf{A}(r,\theta,\phi)$ in spherical coordinates, we must take the scalar product of the operator ∇ and the vector field. This operation involves the evaluation of such derivatives as $\partial\mathbf{i}_r/\partial r$, $\partial\mathbf{i}_r/\partial\theta$, . . . , which is complicated by the changing orientations of these unit vectors from point to point. Consequently, it is easier to calculate the divergence of $\mathbf{A}(r,\theta,\phi)$ through the use of Gauss' theorem (3-61).

Let the vector field $\mathbf{A}(r,\theta,\phi)$ represent the rate of flow per unit area as discussed in Sec. 3-4. The mass dQ_a of fluid flowing per unit time into the spherical volume element shown in Fig. 7-2 through the surface $P_1P_2P_3P_4$ is

$$dQ_a = A_r\, dl_\theta\, dl_\phi \tag{7-16}$$

The mass dQ_b of fluid flowing per unit time out of the volume element through the surface $P_5P_6P_7P_8$ differs from dQ_a for two reasons: (1) The values of the flow rate per unit area A_r differ on the two surfaces and (2) the surface area of $P_5P_6P_7P_8$ is larger than that of $P_1P_2P_3P_4$. Thus, dQ_b may be expanded in a Taylor series,

$$dQ_b = A_r\, dl_\theta\, dl_\phi + \frac{\partial}{\partial l_r}(A_r\, dl_\theta\, dl_\phi)\, dl_r \tag{7-17}$$

where the higher-order terms are neglected. The net outward rate of flow through these two faces is $dQ_b - dQ_a$ or

$$\frac{\partial}{\partial l_r}(A_r\, dl_\theta\, dl_\phi)\, dl_r = \frac{\partial}{\partial r}(r^2 \sin\theta\ A_r)\, dr\, d\theta\, d\phi \tag{7-18}$$

The net outward rate of flow through the two faces normal to $\mathbf{i}_\theta$, namely $P_1P_5P_8P_4$ and $P_2P_6P_7P_3$, is

$$\begin{aligned} A_\theta\, dl_r\, dl_\phi &+ \frac{\partial}{\partial l_\theta}(A_\theta\, dl_r\, dl_\phi)\, dl_\theta - A_\theta\, dl_r\, dl_\phi \\ &= \frac{\partial}{\partial l_\theta}(A_\theta\, dl_r\, dl_\phi)\, dl_\theta \\ &= \frac{\partial}{\partial \theta}(r \sin\theta\ A_\theta)\, dr\, d\theta\, d\phi \end{aligned} \tag{7-19}$$

Similarly, the net outward rate of flow through the faces normal to $\mathbf{i}_\phi$ is

$$\begin{aligned} A_\phi\, dl_r\, dl_\theta &+ \frac{\partial}{\partial l_\phi}(A_\phi\, dl_r\, dl_\theta)\, dl_\phi - A_\phi\, dl_r\, dl_\theta \\ &= \frac{\partial}{\partial l_\phi}(A_\phi\, dl_r\, dl_\theta)\, dl_\phi \\ &= \frac{\partial}{\partial \phi}(rA_\phi)\, dr\, d\theta\, d\phi \end{aligned} \tag{7-20}$$

The net outward rate of flow dQ through the total surface of the volume element is the sum of these three contributions:

$$dQ = \left[\frac{\partial}{\partial r}(r^2 \sin\theta\, A_r) + \frac{\partial}{\partial \theta}(r \sin\theta\, A_\theta) + \frac{\partial}{\partial \phi}(rA_\phi)\right] dr\, d\theta\, d\phi \tag{7-21}$$

Now, the total outward rate of flow Q from a volume V is

$$Q = \int_V dQ = \int_V \nabla \cdot \mathbf{A}\, dv \tag{3-68}$$

Comparing Eqs. (3-68), (7-11), and (7-21), we find that the divergence of **A** in spherical coordinates is given by

$$\nabla \cdot \mathbf{A} = \frac{1}{r^2}\frac{\partial}{\partial r}(r^2 A_r) + \frac{1}{r \sin\theta}\frac{\partial}{\partial \theta}(\sin\theta\, A_\theta) + \frac{1}{r\sin\theta}\frac{\partial A_\phi}{\partial \phi} \tag{7-22}$$

The laplacian of a scalar function $\psi(r,\theta,\phi)$ is obtained from Eq. (7-22) by letting **A** be the gradient of ψ. According to Eq. (7-14), the components of $\mathbf{A}(r,\theta,\phi)$ become

$$A_r = \frac{\partial \psi}{\partial r} \qquad A_\theta = \frac{1}{r}\frac{\partial \psi}{\partial \theta} \qquad A_\phi = \frac{1}{r\sin\theta}\frac{\partial \psi}{\partial \phi} \tag{7-23}$$

Substitution of Eqs. (7-23) into Eq. (7-22) yields

$$\nabla^2\psi = \frac{1}{r^2}\frac{\partial}{\partial r}\left(r^2\frac{\partial \psi}{\partial r}\right) + \frac{1}{r^2 \sin\theta}\frac{\partial}{\partial \theta}\left(\sin\theta\,\frac{\partial \psi}{\partial \theta}\right) + \frac{1}{r^2\sin^2\theta}\frac{\partial^2 \psi}{\partial \phi^2} \tag{7-24}$$

7-5 Curl of a vector field

The curl of a vector function $\mathbf{A}(r,\theta,\phi)$ in spherical coordinates is calculated most easily by means of Stokes' theorem (3-76). We calculate the line integrals of **A** around the edges of each surface of the volume element in Fig. 7-2 and equate each line integral to the respective surface integral of the curl of **A**.

We first consider the surface $P_1P_2P_3P_4$. According to the right-hand rule, we select the path $(P_1P_4P_3P_2P_1)$ to correspond to the outward normal to the surface. The line integrals of **A** along this path are as follows:

Along (P_1P_4): $\quad A_\phi\, dl_\phi$

Along (P_4P_3): $\quad A_\theta\, dl_\theta + \dfrac{\partial}{\partial l_\phi}(A_\theta\, dl_\theta)\, dl_\phi$

Along (P_3P_2): $\quad -[A_\phi\, dl_\phi + \dfrac{\partial}{\partial l_\phi}(A_\phi\, dl_\phi)\, dl_\theta]$

Along (P_2P_1): $\quad -A_\theta\, dl_\theta$

The last two expressions are negative because these portions of the path are traced in the negative θ and ϕ directions. The line integral along this path is, then,

$$(\mathbf{A} \cdot d\mathbf{l})_{P_1P_4P_3P_2} = \frac{\partial}{\partial l_\phi}(A_\theta \, dl_\theta)\, dl_\phi - \frac{\partial}{\partial l_\theta}(A_\phi \, dl_\phi)\, dl_\theta$$

$$= \left[\frac{\partial}{\partial \phi}(rA_\theta) - \frac{\partial}{\partial \theta}(r \sin\theta \, A_\phi)\right] d\theta \, d\phi \tag{7-25}$$

If we equate this line integral to the r component of the surface integral of $\nabla \times \mathbf{A}$, we obtain

$$(\nabla \times \mathbf{A})_r \, ds_r = \left[\frac{\partial}{\partial \phi}(rA_\theta) - \frac{\partial}{\partial \theta}(r \sin\theta \, A_\phi)\right] d\theta \, d\phi \tag{7-26}$$

On the surface $P_1P_2P_3P_4$, the outward normal to the surface is in the negative r direction, so that

$$ds_r = -dl_\theta \, dl_\phi = -r^2 \sin\theta \, d\theta \, d\phi \tag{7-27}$$

Combining Eqs. (7-26) and (7-27), we find that

$$(\nabla \times \mathbf{A})_r = \frac{1}{r \sin\theta}\left[\frac{\partial}{\partial \theta}(\sin\theta \, A_\phi) - \frac{\partial A_\theta}{\partial \phi}\right] \tag{7-28}$$

Keeping terms with only the first derivatives in the Taylor series expansions, we obtain an identical expression for $(\nabla \times \mathbf{A})_r$ for the surface $P_5P_6P_7P_8$.

For the path $(P_1P_5P_8P_4P_1)$, the line integral is

$$(\mathbf{A} \cdot d\mathbf{l})_{P_1P_5P_8P_4} = \frac{\partial}{\partial l_r}(A_\phi \, dl_\phi)\, dl_r - \frac{\partial}{\partial l_\phi}(A_r \, dl_r)\, dl_\phi$$

$$= \left[\frac{\partial}{\partial r}(r \sin\theta \, A_\phi) - \frac{\partial A_r}{\partial \phi}\right] dr \, d\phi \tag{7-29}$$

Equating this expression to the θ component of $(\nabla \times \mathbf{A}) \cdot d\mathbf{s}$ and noting that for this surface $ds_\theta = -dl_r \, dl_\phi$, we obtain

$$(\nabla \times \mathbf{A})_\theta = \frac{1}{r}\left[\frac{1}{\sin\theta}\frac{\partial A_r}{\partial \phi} - \frac{\partial}{\partial r}(rA_\phi)\right] \tag{7-30}$$

The same result may be obtained from the surface $P_2P_6P_7P_3$.

For the path $(P_2P_6P_5P_1P_2)$ and for the path $(P_3P_4P_8P_7P_3)$, Stokes' theorem may be used to obtain the ϕ component of $\nabla \times \mathbf{A}$:

$$(\nabla \times \mathbf{A})_\phi = \frac{1}{r}\left[\frac{\partial}{\partial r}(rA_\theta) - \frac{\partial A_r}{\partial \theta}\right] \tag{7-31}$$

Equations (7-28), (7-30), and (7-31) can be expressed in the form of a determinant:

$$\nabla \times \mathbf{A} = \frac{1}{r^2 \sin\theta} \begin{vmatrix} \mathbf{i}_r & r\mathbf{i}_\theta & r\sin\theta\ \mathbf{i}_\phi \\ \dfrac{\partial}{\partial r} & \dfrac{\partial}{\partial \theta} & \dfrac{\partial}{\partial \phi} \\ A_r & rA_\theta & r\sin\theta\ A_\phi \end{vmatrix} \tag{7-32}$$

PROBLEMS

1. If a vector field **A** is given in spherical coordinates by the expression

   $\mathbf{A} = \mathbf{i}_r r + \mathbf{i}_\theta \sec\theta + \mathbf{i}_\phi r^3 \sin\theta$

   express **A** in cartesian coordinates.

2. If a central field **A** is defined by $\mathbf{A} = \mathbf{i}_r f(r)$, determine $f(r)$ so that the field is solenoidal. Show that **A** is also irrotational and find the scalar potential ψ such that $\mathbf{A} = \nabla\psi$.

3. A scalar function ψ in spherical coordinates is given by

   $\psi = (ar^3 + br^{-2}) \cos 2\theta \ \sin\phi$

   (*a*) Find $\mathbf{A} = \nabla\psi$.

   (*b*) Find $\nabla \times \mathbf{A}$.

   (*c*) Find $\nabla^2\psi$.

4. A vector field **A** is given by

   $\mathbf{A} = \mathbf{i}_r e^{-r} \sin\theta + \mathbf{i}_\theta r^{-1} \sin\theta + \mathbf{i}_\phi \cos\phi$

   Find $\nabla \cdot \mathbf{A}$ and $\nabla \times \mathbf{A}$.

5. A vector field is given by $\mathbf{A} = \mathbf{i}_r r^2$.

   (*a*) Evaluate the volume integral $\int_V \nabla \cdot \mathbf{A}\, dv$ throughout the octant of a sphere bounded by $r = a$ and the planes $x = 0, y = 0, z = 0$.

   (*b*) Evaluate the surface integral $\oint_S \mathbf{A} \cdot d\mathbf{s}$ over the surface of the octant and verify Gauss' theorem for this example.

6. Given the vector field **A**

   $\mathbf{A} = \mathbf{i}_\theta \cos\left(\frac{3\theta}{2}\right) + \mathbf{i}_\phi r \sin\left(\frac{3\theta}{2}\right)$

(*a*) Evaluate the surface integral $\int_S (\nabla \times \mathbf{A}) \cdot d\mathbf{s}$ over a circle of radius a, center at the origin, in the $\theta = \pi/2$ plane (the xy plane).

(*b*) Evaluate the line integral $\oint \mathbf{A} \cdot d\mathbf{l}$ around the circle and verify Stokes' theorem for this example.

7. A sphere of radius a has an electric charge density proportional to the distance r from the center (that is, $\rho = Cr$).

(*a*) Using the Maxwell equation (4-12) and the constitutive equation (4-25), find the electric field vector **E** inside and outside the sphere.

(*b*) Obtain expressions for the scalar potential in Eq. (4-112) inside and outside the sphere.

(*c*) Evaluate $\nabla \times \mathbf{E}$ inside and outside the sphere.

Chapter eight

Orthogonal Curvilinear Coordinates

8-1 Orthogonal coordinate systems

In Chap. 7 we showed that vectors may be expressed in a coordinate system other than a cartesian one by developing in detail vector analysis in terms of spherical coordinates. In this chapter we consider vectors in an arbitrary orthogonal coordinate system, of which the cartesian and spherical systems are just special cases.

The three coordinates of a point in space are denoted by q_1, q_2, q_3. These coordinates are related to the cartesian coordinates x, y, z by

$$\begin{aligned} x &= x(q_1,q_2,q_3) \\ y &= y(q_1,q_2,q_3) \\ z &= z(q_1,q_2,q_3) \end{aligned} \tag{8-1}$$

The surfaces q_1 = constant, q_2 = constant, q_3 = constant are called *coordinate surfaces,* which we assume are mutually orthogonal. Any two coordinate surfaces intersect in a *coordinate line*; three coordinate surfaces intersect at a point. Thus, simultaneous specification of the three surfaces q_1 = constant, q_2 = constant, q_3 = constant determines the location of the point P in this coordinate system.

We next introduce a set of unit vectors $\mathbf{i}_1, \mathbf{i}_2, \mathbf{i}_3$, corresponding to q_1, q_2, q_3, respectively. Each unit vector is tangent to a coordinate line at a point P. The vector $\mathbf{i}_1$ is tangent to the coordinate line determined by q_2 = constant and q_3 = constant. Similarly, $\mathbf{i}_2$ and $\mathbf{i}_3$ are tangent to the coordinate lines determined by q_1= constant, q_3 = constant and by q_1 = constant, q_2 = constant, respectively. By virtue of the orthogonality of the coordinate surfaces, these three unit vectors are mutually perpendicular at any given point P. A *right-handed coordinate system* is one in which the unit vectors at P are related by

$$\mathbf{i}_1 \times \mathbf{i}_2 = \mathbf{i}_3 \tag{8-2}$$

As discussed in Sec. 7-1 for the spherical coordinate system, the unit vectors $\mathbf{i}_1, \mathbf{i}_2, \mathbf{i}_3$ change direction from point to point in space. However, at any given point they are mutually perpendicular and Eq. (8-2) applies.

A vector $\mathbf{A}$ may be expressed in terms of its components in the q_1, q_2, q_3 coordinate system,

$$\mathbf{A} = \mathbf{i}_1 A_1 + \mathbf{i}_2 A_2 + \mathbf{i}_3 A_3 \tag{8-3}$$

where A_1, A_2, A_3 are its components in that coordinate system. Since, in general, the unit vectors $\mathbf{i}_1, \mathbf{i}_2, \mathbf{i}_3$ change direction from point to point in space, it is not possible to add or subtract vectors or to form scalar products and vector products directly. It is necessary to transform the vectors and the unit vectors to cartesian coordinates through the use of Eqs. (8-1) as was illustrated for spherical coordinates in Sec. 7-1.

8-2 Volume element

A volume element dv in a generalized orthogonal coordinate system is shown in Fig. 8-1, where $P_1, \ldots, P_8$ denote the corners of the element. The drawing is, of course, stylized and represents no particular coordinate system. The faces are determined as follows:

$$\begin{aligned}
&P_1P_2P_3P_4: & q_1 &= \text{const}\\
&P_5P_6P_7P_8: & q_1 + dq_1 &= \text{const}\\
&P_1P_5P_8P_4: & q_2 &= \text{const}\\
&P_2P_6P_7P_3: & q_2 + dq_2 &= \text{const}\\
&P_1P_2P_6P_5: & q_3 &= \text{const}\\
&P_4P_3P_7P_8: & q_3 + dq_3 &= \text{const}
\end{aligned}$$

The edges (P_1P_5), (P_1P_2), and (P_1P_4) are denoted by dl_1, dl_2, dl_3, respectively. Since each arc length is proportional to the differential change in the corresponding generalized coordinate, we write

$$dl_1 = h_1 dq_1 \qquad dl_2 = h_2 dq_2 \qquad dl_3 = h_3 dq_3 \tag{8-4}$$

where h_1, h_2, h_3 are scale factors. The arc lengths dl_1, dl_2, dl_3 must have the dimensions of length, but q_1, q_2, q_3 need not.

In the limit $dq_1 \to 0, dq_2 \to 0, dq_3 \to 0$, the arc lengths approach straight lines, so that the volume of the element is

$$dv = dl_1\, dl_2\, dl_3 = h_1 h_2 h_3\, dq_1\, dq_2\, dq_3 \tag{8-5}$$

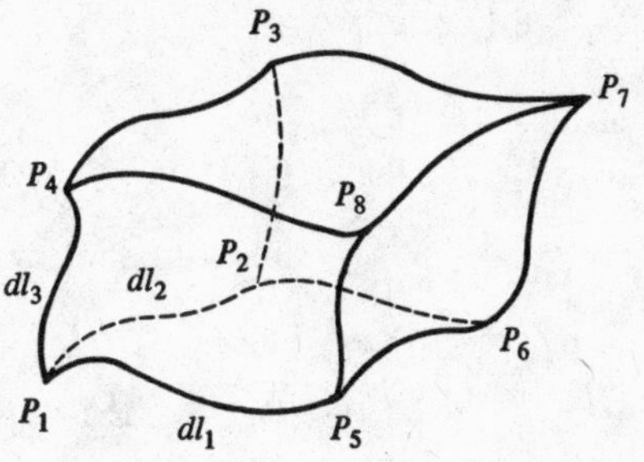

FIGURE 8-1

A useful device for finding the volume element in an arbitrary coordinate system is the *jacobian determinant* J. The elements J_{ij} of the determinant are given by

$$J_{ij} = \frac{\partial x_i}{\partial q_j} \qquad i, j = 1, 2, 3 \tag{8-6}$$

where, as before, $x_1 = x$, $x_2 = y$, $x_3 = z$. The product $h_1 h_2 h_3$ is, then, the absolute value of the jacobian determinant J, so that

$$dv = |J| \, dq_1 \, dq_2 \, dq_3 \tag{8-7}$$

8-3 Scale factors

The differential vector distance $d\mathbf{l}$ between two nearby points may be expressed in generalized orthogonal coordinates by

$$d\mathbf{l} = \mathbf{i}_1 dl_1 + \mathbf{i}_2 dl_2 + \mathbf{i}_3 dl_3 \tag{8-8}$$

Note that $d\mathbf{l}$ is the diagonal of the volume element in Fig. 8-1 and the two nearby points in question are P_1 and P_7. The square of the magnitude of $d\mathbf{l}$ is just $dl \cdot dl$. Since $\mathbf{i}_1, \mathbf{i}_2, \mathbf{i}_3$ are mutually orthogonal, we have

$$dl^2 = dl_1^2 + dl_2^2 + dl_3^2 \tag{8-9}$$

or

$$dl^2 = h_1^2 \, dq_1^2 + h_2^2 \, dq_2^2 + h_3^2 \, dq_3^2 \tag{8-10}$$

where Eqs. (8-4) have been introduced.

The square of the distance between two nearby points is given in cartesian coordinates by

$$dl^2 = dx^2 + dy^2 + dz^2 = \sum_{i=1}^{3} dx_i^2 \tag{8-11}$$

The three cartesian coordinates x_i are related to the generalized coordinates q_i by

Eqs. (8-1), so that

$$dx_i = \sum_{j=1}^{3} \frac{\partial x_i}{\partial q_j} dq_j \tag{8-12}$$

When Eq. (8-12) is substituted into Eq. (8-11), we obtain

$$\begin{aligned} dl^2 &= Q_{11}\, dq_1^2 + Q_{22}\, dq_2^2 + Q_{33}\, dq_3^2 \\ &\quad + 2Q_{12}\, dq_1\, dq_2 + 2Q_{13}\, dq_1\, dq_3 + 2Q_{23}\, dq_2\, dq_3 \end{aligned} \tag{8-13}$$

where

$$Q_{ij} = \frac{\partial x}{\partial q_i}\frac{\partial x}{\partial q_j} + \frac{\partial y}{\partial q_i}\frac{\partial y}{\partial q_j} + \frac{\partial z}{\partial q_i}\frac{\partial z}{\partial q_j} \tag{8-14}$$

Now, the distance dl between two points must be the same for all coordinate systems, so that dl^2 in Eq. (8-13) must equal dl^2 in Eq. (8-10). Therefore, we obtain the requirement for an orthogonal coordinate system that

$$\begin{aligned} Q_{ii} &= h_i^2 \qquad i = 1, 2, 3 \\ Q_{ij} &= 0 \qquad i \neq j \end{aligned} \tag{8-15}$$

Thus, the scale factors h_i are readily determined from Eqs. (8-1) by the relation

$$h_i = \left[\left(\frac{\partial x}{\partial q_i}\right)^2 + \left(\frac{\partial y}{\partial q_i}\right)^2 + \left(\frac{\partial z}{\partial q_i}\right)^2\right]^{1/2} \qquad i = 1, 2, 3 \tag{8-16}$$

For some coordinate systems, such as the spherical system discussed in Chap. 7, the scalar factors are also easily determined from geometrical considerations.

8-4 Gradient of a scalar field

If $\psi(q_1, q_2, q_3)$ is a continuous and differentiable scalar function of position in space, its space derivatives in a generalized orthogonal coordinate system are $\partial\psi/\partial l_1$, $\partial\psi/\partial l_2$, $\partial\psi/\partial l_3$. The gradient of $\psi(q_1, q_2, q_3)$ is a vector in this coordinate system with these space derivatives as components:

$$\begin{aligned} \nabla\psi &= \mathbf{i}_1 \frac{\partial\psi}{\partial l_1} + \mathbf{i}_2 \frac{\partial\psi}{\partial l_2} + \mathbf{i}_3 \frac{\partial\psi}{\partial l_3} \\ &= \mathbf{i}_1 \frac{1}{h_1}\frac{\partial\psi}{\partial q_1} + \mathbf{i}_2 \frac{1}{h_2}\frac{\partial\psi}{\partial q_2} + \mathbf{i}_3 \frac{1}{h_3}\frac{\partial\psi}{\partial q_3} \end{aligned} \tag{8-17}$$

Thus, the vector differential operator ∇ has the form

$$\nabla = \mathbf{i}_1 \frac{1}{h_1}\frac{\partial}{\partial q_1} + \mathbf{i}_2 \frac{1}{h_2}\frac{\partial}{\partial q_2} + \mathbf{i}_3 \frac{1}{h_3}\frac{\partial}{\partial q_3} \tag{8-18}$$

8-5 Divergence of a vector field

To calculate the divergence of a vector field $\mathbf{A}(q_1,q_2,q_3)$, we use Gauss' theorem (3-61) and follow a procedure parallel to the one described in Sec. 7-4 for spherical coordinates.

The contribution to the surface integral $\int_S \mathbf{A} \cdot d\mathbf{s}$ of the face $P_1P_2P_3P_4$ in Fig. 8-1 is

$$-A_1\,dl_2\,dl_3$$

and of the face $P_5P_6P_7P_8$ is

$$A_1\,dl_2\,dl_3 + \frac{\partial}{\partial l_1}(A_1\,dl_2\,dl_3)\,dl_1$$

Thus, the total contribution of these two faces is

$$\frac{\partial}{\partial l_1}(A_1\,dl_2\,dl_3)\,dl_1 = \frac{\partial}{\partial q_1}(A_1 h_2 h_3)\,dq_1\,dq_2\,dq_3$$

$$= \frac{1}{h_1 h_2 h_3}\frac{\partial}{\partial q_1}(A_1 h_2 h_3)\,dv \qquad (8\text{-}19)$$

Adding the corresponding expressions for the other two pairs of faces, integrating, and comparing the result with Gauss' theorem (3-61) give

$$\nabla \cdot \mathbf{A} = \frac{1}{h_1 h_2 h_3}\left[\frac{\partial}{\partial q_1}(A_1 h_2 h_3) + \frac{\partial}{\partial q_2}(A_2 h_1 h_3) + \frac{\partial}{\partial q_3}(A_3 h_1 h_2)\right] \qquad (8\text{-}20)$$

Replacing $\mathbf{A}$ in Eq. (8-20) by $\nabla\psi$ yields an expression for the laplacian of a scalar function $\psi(q_1,q_2,q_3)$. According to Eq. (8-17), the components of $\mathbf{A}$ become

$$A_1 = \frac{1}{h_1}\frac{\partial\psi}{\partial q_1} \qquad A_2 = \frac{1}{h_2}\frac{\partial\psi}{\partial q_2} \qquad A_3 = \frac{1}{h_3}\frac{\partial\psi}{\partial q_3} \qquad (8\text{-}21)$$

Thus, the laplacian of ψ is

$$\nabla^2\psi = \frac{1}{h_1 h_2 h_3}\left[\frac{\partial}{\partial q_1}\left(\frac{h_2 h_3}{h_1}\frac{\partial\psi}{\partial q_1}\right) + \frac{\partial}{\partial q_2}\left(\frac{h_1 h_3}{h_2}\frac{\partial\psi}{\partial q_2}\right) + \frac{\partial}{\partial q_3}\left(\frac{h_1 h_2}{h_3}\frac{\partial\psi}{\partial q_3}\right)\right] \qquad (8\text{-}22)$$

8-6 Curl of a vector field

The curl of a vector function $\mathbf{A}(q_1,q_2,q_3)$ in generalized orthogonal coordinates is obtained from Stokes' theorem (3-76). We follow here the procedure used in Sec. 7-5 for the curl of a vector field in spherical coordinates.

Component 1 of $\nabla \times \mathbf{A}$ is obtained by applying Stokes' theorem to the surface $P_1P_2P_3P_4$ in Fig. 8-1. According to the right-hand rule, the path $(P_1P_4P_3P_2P_1)$ corresponds to the outward normal to the surface. The line integrals are as follows:

Along (P_1P_4): $\quad A_3\,dl_3$

Along (P_4P_3): $\quad A_2\,dl_2 + \dfrac{\partial}{\partial l_3}(A_2\,dl_2)\,dl_3$

Along (P_3P_2): $\quad -\left[A_3\,dl_3 + \dfrac{\partial}{\partial l_2}(A_3\,dl_3)\,dl_2\right]$

Along (P_2P_1): $\quad -A_2\,dl_2$

so that

$$(\mathbf{A}\cdot d\mathbf{l})_{P_1P_4P_3P_2} = \frac{\partial}{\partial l_3}(A_2\,dl_2)\,dl_3 - \frac{\partial}{\partial l_2}(A_3\,dl_3)\,dl_2$$

$$= \left[\frac{\partial}{\partial q_3}(h_2A_2) - \frac{\partial}{\partial q_2}(h_3A_3)\right] dq_2\,dq_3 \tag{8-23}$$

By Stokes' theorem this line integral is equal to component 1 of $\nabla \times \mathbf{A}$ multiplied by the area of the surface $P_1P_2P_3P_4$:

$$(\nabla \times \mathbf{A})_1\,ds_1 = \left[\frac{\partial}{\partial q_3}(h_2A_2) - \frac{\partial}{\partial q_2}(h_3A_3)\right] dq_2\,dq_3 \tag{8-24}$$

Since

$$ds_1 = -dl_2\,dl_3 = -h_2h_3\,dq_2\,dq_3 \tag{8-25}$$

Eq. (8-24) becomes

$$(\nabla \times \mathbf{A})_1 = \frac{1}{h_2h_3}\left[\frac{\partial}{\partial q_2}(h_3A_3) - \frac{\partial}{\partial q_3}(h_2A_2)\right] \tag{8-26}$$

Similar expressions may be obtained for components 2 and 3 of the curl of **A**. The final result is

$$\nabla \times \mathbf{A} = \frac{1}{h_1h_2h_3}\begin{vmatrix} h_1\mathbf{i}_1 & h_2\mathbf{i}_2 & h_3\mathbf{i}_3 \\ \dfrac{\partial}{\partial q_1} & \dfrac{\partial}{\partial q_2} & \dfrac{\partial}{\partial q_3} \\ h_1A_1 & h_2A_2 & h_3A_3 \end{vmatrix}$$

$$= \mathbf{i}_1\frac{1}{h_2h_3}\left[\frac{\partial}{\partial q_2}(h_3A_3) - \frac{\partial}{\partial q_3}(h_2A_2)\right] + \mathbf{i}_2\frac{1}{h_1h_3}\left[\frac{\partial}{\partial q_3}(h_1A_1) - \frac{\partial}{\partial q_1}(h_3A_3)\right] + \mathbf{i}_3\frac{1}{h_1h_2}\left[\frac{\partial}{\partial q_1}(h_2A_2) - \frac{\partial}{\partial q_2}(h_1A_1)\right] \tag{8-27}$$

8-7 Cylindrical coordinates

The cylindrical coordinates ρ, ϕ, z, which are related to the cartesian coordinates

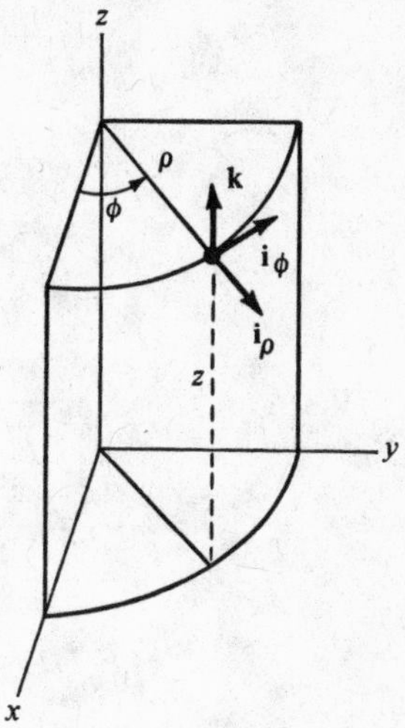

FIGURE 8-2

x, y, z by

$$x = \rho \cos \phi$$
$$y = \rho \sin \phi \qquad (8\text{-}28)$$
$$z = z$$

constitute an orthogonal coordinate system which is frequently employed. In this system of coordinates, an arbitrary vector **A** may be written in component form as

$$\mathbf{A} = \mathbf{i}_\rho A_\rho + \mathbf{i}_\phi A_\phi + \mathbf{k} A_z \qquad (8\text{-}29)$$

where $\mathbf{i}_\rho$, $\mathbf{i}_\phi$, **k** are the unit vectors which are shown in Fig. 8-2 and which satisfy the criterion

$$\mathbf{i}_\rho \times \mathbf{i}_\phi = \mathbf{k} \qquad (8\text{-}30)$$

From Eqs. (8-16) and (8-28), the scale factors h_ρ, h_ϕ, h_z are easily determined:

$$h_\rho = 1 \qquad h_\phi = \rho \qquad h_z = 1 \qquad (8\text{-}31)$$

The volume element dv is, then,

$$dv = \rho \, d\rho \, d\phi \, dz \qquad (8\text{-}32)$$

The gradient, divergence, laplacian, and curl for cylindrical coordinates are

$$\nabla\psi = \mathbf{i}_\rho \frac{\partial\psi}{\partial\rho} + \mathbf{i}_\phi \frac{1}{\rho}\frac{\partial\psi}{\partial\phi} + \mathbf{k}\frac{\partial\psi}{\partial z} \tag{8-33}$$

$$\nabla \cdot \mathbf{A} = \frac{1}{\rho}\frac{\partial}{\partial\rho}(\rho A_\rho) + \frac{1}{\rho}\frac{\partial A_\phi}{\partial\phi} + \frac{\partial A_z}{\partial z} \tag{8-34}$$

$$\nabla^2\psi = \frac{1}{\rho}\frac{\partial}{\partial\rho}\left(\rho\frac{\partial\psi}{\partial\rho}\right) + \frac{1}{\rho^2}\frac{\partial^2\psi}{\partial\phi^2} + \frac{\partial^2\psi}{\partial z^2} \tag{8-35}$$

$$(\nabla \times \mathbf{A})_\rho = \frac{1}{\rho}\frac{\partial A_z}{\partial\phi} - \frac{\partial A_\phi}{\partial z} \tag{8-36a}$$

$$(\nabla \times \mathbf{A})_\phi = \frac{\partial A_\rho}{\partial z} - \frac{\partial A_z}{\partial\rho} \tag{8-36b}$$

$$(\nabla \times \mathbf{A})_z = \frac{1}{\rho}\left[\frac{\partial}{\partial\rho}(\rho A_\phi) - \frac{\partial A_\rho}{\partial\phi}\right] \tag{8-36c}$$

8-8 Other coordinate systems

Although cartesian, spherical, and cylindrical coordinates are by far the most frequently used coordinate systems, certain applications of vector analysis require other choices. Accordingly, we list here a number of other orthogonal coordinate systems along with their equations of transformation in terms of cartesian coordinates and their scale factors. The volume element, gradient, divergence, laplacian, and curl may be readily determined by means of the general formulas derived in this chapter. A more detailed discussion of these and other orthogonal coordinate systems may be found in the book by Margenau and Murphy.[1]

Parabolic coordinates:

$$x = \sqrt{\xi\eta}\cos\varphi$$
$$y = \sqrt{\xi\eta}\sin\varphi$$
$$z = \tfrac{1}{2}(\xi - \eta)$$

$$h_\xi = \frac{1}{2}\sqrt{\frac{\xi+\eta}{\xi}} \qquad h_\eta = \frac{1}{2}\sqrt{\frac{\xi+\eta}{\eta}} \qquad h_\varphi = \sqrt{\xi\eta}$$

[1] H. Margenau and G. M. Murphy, "The Mathematics of Physics and Chemistry," 2d ed., Chap. 5, D. Van Nostrand Company, Inc., Princeton, N.J., 1956.

Confocal elliptic coordinates (prolate spheroids):

$$x = a\sqrt{\xi^2 - 1}\sqrt{1 - \eta^2}\cos\varphi$$
$$y = a\sqrt{\xi^2 - 1}\sqrt{1 - \eta^2}\sin\varphi$$
$$z = a\xi\eta$$

In terms of the distances r_A and r_B from the points $(0, 0, -a)$ and $(0, 0, a)$, respectively, ξ and η are given by the expressions

$$\xi = \frac{r_A + r_B}{2a} \qquad \eta = \frac{r_A - r_B}{2a}$$

$$h_\xi = a\sqrt{\frac{\xi^2 - \eta^2}{\xi^2 - 1}} \qquad h_\eta = a\sqrt{\frac{\xi^2 - \eta^2}{1 - \eta^2}} \qquad h_\varphi = a\sqrt{(\xi^2 - 1)(1 - \eta^2)}$$

Spheroidal coordinates (oblate spheroids):

$$x = a\xi\eta\cos\varphi \qquad y = a\xi\eta\sin\varphi \qquad z = a\sqrt{(\xi^2 - 1)(1 - \eta^2)}$$

$$h_\xi = a\sqrt{\frac{\xi^2 - \eta^2}{\xi^2 - 1}} \qquad h_\eta = a\sqrt{\frac{\xi^2 - \eta^2}{1 - \eta^2}} \qquad h_\varphi = a\xi\eta$$

Parabolic cylindrical coordinates:

$$x = \frac{1}{2}(u - v) \qquad y = \sqrt{uv} \qquad z = z$$

$$h_u = \frac{1}{2}\sqrt{\frac{u + v}{u}} \qquad h_v = \frac{1}{2}\sqrt{\frac{u + v}{v}} \qquad h_z = 1$$

Elliptic cylindrical coordinates:

$$x = a\sqrt{(u^2 - 1)(1 - v^2)} \qquad y = auv \qquad z = z$$

$$h_u = a\sqrt{\frac{u^2 - v^2}{u^2 - 1}} \qquad h_v = a\sqrt{\frac{u^2 - v^2}{1 - v^2}} \qquad h_z = 1$$

Ellipsoidal coordinates:

$$x^2 = \frac{(a^2 + u)(a^2 + v)(a^2 + w)}{(a^2 - b^2)(a^2 - c^2)} \qquad y^2 = \frac{(b^2 + u)(b^2 + v)(b^2 + w)}{(b^2 - c^2)(b^2 - a^2)}.$$

$$z^2 = \frac{(c^2 + u)(c^2 + v)(c^2 + w)}{(c^2 - a^2)(c^2 - b^2)}$$

$$h_u^2 = \frac{(u - v)(u - w)}{4(a^2 + u)(b^2 + u)(c^2 + u)} \qquad h_v^2 = \frac{(v - w)(v - u)}{4(a^2 + v)(b^2 + v)(c^2 + v)}$$

$$h_w^2 = \frac{(w - u)(w - v)}{4(a^2 + w)(b^2 + w)(c^2 + w)}$$

Confocal parabolic coordinates:

$$x = \tfrac{1}{2}(u + v + w - a - b) \qquad y^2 = \frac{(a-u)(a-v)(a-w)}{b-a}$$

$$z^2 = \frac{(b-u)(b-v)(b-w)}{a-b} \qquad u > b > v > a > w$$

$$h_u^2 = \frac{(u-v)(u-w)}{4(a-u)(b-u)} \qquad h_v^2 = \frac{(v-u)(v-w)}{4(a-v)(b-v)} \qquad h_w^2 = \frac{(w-u)(w-v)}{4(a-w)(b-w)}$$

PROBLEMS

1. A scalar function ψ is given in cylindrical coordinates by

$$\psi = (1 - \rho^{-2})\rho \sin\phi$$

Find $\nabla\psi$ and $\nabla^2\psi$ in both cylindrical and cartesian coordinates.

2. A vector field **A** is given in cylindrical coordinates by

$$\mathbf{A} = \mathbf{i}_\rho\, \rho z \sin\phi + \mathbf{i}_\phi\, 2\rho z \cos\phi + \mathbf{k}z^2$$

Find $\nabla \cdot \mathbf{A}$ and $\nabla \times \mathbf{A}$. Verify that $\nabla \cdot (\nabla \times \mathbf{A}) = 0$.

3. A vector field in cylindrical coordinates is

$$\mathbf{A} = \mathbf{i}_\rho \rho \cos^2\phi + \mathbf{i}_\phi \rho \sin^2\phi$$

(*a*) Evaluate $\nabla \cdot \mathbf{A}$.

(*b*) Evaluate the surface integral of **A** over the surface of a quadrant of a cylinder with unit height whose cross section is bounded by the positive x and y axes and the arc of the circle $x^2 + y^2 = a^2$.

(*c*) Verify Gauss' theorem for this example.

4. A vector field **A** in cylindrical coordinates is given by

$$\mathbf{A} = \mathbf{i}_\rho \rho^2 \cos\phi + \mathbf{i}_\phi \rho^2 \sin\phi$$

(*a*) Evaluate the surface integral $\int_S (\nabla \times \mathbf{A}) \cdot d\mathbf{s}$ over the planar surface bounded by the positive x and y axes and the quadrant of a circle of radius a with center at the origin.

(*b*) Evaluate the line integral $\oint \mathbf{A} \cdot d\mathbf{l}$ over the perimeter of this surface and verify Stokes' theorem for this example.

5. An infinitely long nonconducting cylinder has an electric charge density equal to $C\rho^{-1/2}$.

(*a*) Using the Maxwell equation (4-12) and the constitutive equation (4-25), find the electric field vector **E** at points inside and outside the cylinder.

(*b*) Obtain expressions for the scalar potential in Eq. (4-112) inside and outside the cylinder.

(*c*) Evaluate $\nabla \times \mathbf{E}$ inside and outside the cylinder.

(Throughout this problem be careful to distinguish between the charge density and the radial distance, both of which are given the symbol ρ, and between the scalar potential and the axial angle, both of which are labeled ϕ.)

6. Show that the laplacian in parabolic coordinates is

$$\nabla^2 = \frac{4}{\xi + \eta}\left[\frac{\partial}{\partial \xi}\left(\xi\frac{\partial}{\partial \xi}\right) + \frac{\partial}{\partial \eta}\left(\eta\frac{\partial}{\partial \eta}\right)\right] + \frac{1}{\xi\eta}\frac{\partial^2}{\partial \phi^2}$$

7. Show that the laplacian in confocal elliptic coordinates is

$$\nabla^2 = \frac{1}{a^2(\xi^2 - \eta^2)}\left\{\frac{\partial}{\partial \xi}\left[(\xi^2 - 1)\frac{\partial}{\partial \xi}\right] + \frac{\partial}{\partial \eta}\left[(1 - \eta^2)\frac{\partial}{\partial \eta}\right] + \frac{\xi^2 - \eta^2}{(\xi^2 - 1)(1 - \eta^2)}\frac{\partial^2}{\partial \phi^2}\right\}$$

Appendix Summary of Vector and Dyadic Relations

Vector and dyadic operations

$$\mathbf{A}\cdot\mathbf{B}\times\mathbf{C}=\mathbf{B}\cdot\mathbf{C}\times\mathbf{A}=\mathbf{C}\cdot\mathbf{A}\times\mathbf{B}$$

$$\mathbf{A}\times(\mathbf{B}\times\mathbf{C})=(\mathbf{A}\cdot\mathbf{C})\,\mathbf{B}-(\mathbf{A}\cdot\mathbf{B})\,\mathbf{C}$$

$$(\mathbf{A}\times\mathbf{B})\cdot(\mathbf{C}\times\mathbf{D})=(\mathbf{A}\cdot\mathbf{C})(\mathbf{B}\cdot\mathbf{D})-(\mathbf{A}\cdot\mathbf{D})(\mathbf{B}\cdot\mathbf{C})$$

$$(\mathbf{A}\times\mathbf{B})\times(\mathbf{C}\times\mathbf{D})=(\mathbf{A}\times\mathbf{B}\cdot\mathbf{D})\,\mathbf{C}-(\mathbf{A}\times\mathbf{B}\cdot\mathbf{C})\,\mathbf{D}$$

$$(\mathbf{AB})\cdot(\mathbf{CD})=(\mathbf{B}\cdot\mathbf{C})\,\mathbf{AD}$$

$$\mathbf{S}:\mathbf{T}=\mathbf{T}:\mathbf{S}=\sum_{i=1}^{3}\sum_{j=1}^{3}S_{ij}\,T_{ji}$$

$$\mathbf{D}\cdot(\mathbf{AB})\cdot\mathbf{C}=(\mathbf{D}\cdot\mathbf{A})(\mathbf{B}\cdot\mathbf{C})=\mathbf{AB}:\mathbf{CD}=\mathbf{AC}:\mathbf{BD}$$

Differential operations

$$\nabla\cdot(\phi\mathbf{A})=\phi\nabla\cdot\mathbf{A}+\mathbf{A}\cdot\nabla\phi$$

$$\nabla\times(\phi\mathbf{A})=\phi\nabla\times\mathbf{A}+\nabla\phi\times\mathbf{A}$$

$$\nabla(\mathbf{A}\cdot\mathbf{B})=\mathbf{A}\cdot\nabla\mathbf{B}+\mathbf{B}\cdot\nabla\mathbf{A}+\mathbf{A}\times(\nabla\times\mathbf{B})+\mathbf{B}\times(\nabla\times\mathbf{A})$$

$$\nabla\cdot(\mathbf{A}\times\mathbf{B})=\mathbf{B}\cdot(\nabla\times\mathbf{A})-\mathbf{A}\cdot(\nabla\times\mathbf{B})$$

$$\nabla\times(\mathbf{A}\times\mathbf{B})=\mathbf{A}(\nabla\cdot\mathbf{B})-\mathbf{B}(\nabla\cdot\mathbf{A})+\mathbf{B}\cdot\nabla\mathbf{A}-\mathbf{A}\cdot\nabla\mathbf{B}$$

$$\nabla\times\nabla\phi=0$$

$$\nabla\cdot(\nabla\times\mathbf{A})=0$$

$$\nabla\times(\nabla\times\mathbf{A})=\nabla(\nabla\cdot\mathbf{A})-\nabla^2\mathbf{A}$$

$$\nabla \cdot \mathbf{A} = \mathbf{1}{:}\nabla\mathbf{A} = \mathrm{Tr}\,\nabla\mathbf{A}$$

$$\nabla \cdot (\mathbf{AB}) = \mathbf{B}\,\nabla \cdot \mathbf{A} + \mathbf{A} \cdot \nabla\mathbf{B}$$

Gauss' theorem

$$\int_V \nabla \cdot \mathbf{A}\, dv = \oint_S \mathbf{A} \cdot d\mathbf{s}$$

Stokes' theorem

$$\int_S (\nabla \times \mathbf{A}) \cdot d\mathbf{s} = \oint_C \mathbf{A} \cdot d\mathbf{l}$$

Green's theorem

$$\int_V (\phi\nabla^2\psi + \nabla\phi \cdot \nabla\psi)\, dv = \oint_S (\phi\nabla\psi) \cdot d\mathbf{s}$$

$$\int_V (\phi\nabla^2\psi - \psi\nabla^2\phi)\, dv = \oint_S (\phi\nabla\psi - \psi\nabla\phi) \cdot d\mathbf{s}$$

Spherical coordinates (r, θ, ϕ)

$$x = r \sin\theta \cos\phi$$

$$y = r \sin\theta \sin\phi$$

$$z = r \cos\theta$$

$$dv = r^2 \sin\theta\, dr\, d\theta\, d\phi$$

$$\nabla\psi = \mathbf{i}_r \frac{\partial\psi}{\partial r} + \mathbf{i}_\theta \frac{1}{r}\frac{\partial\psi}{\partial\theta} + \mathbf{i}_\phi \frac{1}{r\sin\theta}\frac{\partial\psi}{\partial\phi}$$

$$\nabla^2\psi = \frac{1}{r^2}\frac{\partial}{\partial r}\left(r^2\frac{\partial\psi}{\partial r}\right) + \frac{1}{r^2\sin\theta}\frac{\partial}{\partial\theta}\left(\sin\theta\,\frac{\partial\psi}{\partial\theta}\right) + \frac{1}{r^2\sin^2\theta}\frac{\partial^2\psi}{\partial\phi^2}$$

$$\nabla \cdot \mathbf{A} = \frac{1}{r^2}\frac{\partial}{\partial r}(r^2 A_r) + \frac{1}{r\sin\theta}\frac{\partial}{\partial\theta}(\sin\theta\, A_\theta) + \frac{1}{r\sin\theta}\frac{\partial A_\phi}{\partial\phi}$$

$$(\nabla\times\mathbf{A})_r = \frac{1}{r\sin\theta}\left[\frac{\partial}{\partial\theta}(\sin\theta\, A_\phi) - \frac{\partial A_\theta}{\partial\phi}\right]$$

$$(\nabla\times\mathbf{A})_\theta = \frac{1}{r\sin\theta}\frac{\partial A_r}{\partial\phi} - \frac{1}{r}\frac{\partial}{\partial r}(rA_\phi)$$

$$(\nabla\times\mathbf{A})_\phi = \frac{1}{r}\left[\frac{\partial}{\partial r}(rA_\theta) - \frac{\partial A_r}{\partial\theta}\right]$$

Cylindrical coordinates (ρ, ϕ, z)

$$x = \rho \cos \phi$$

$$y = \rho \sin \phi$$

$$z = z$$

$$dv = \rho \, d\rho \, d\phi \, dz$$

$$\nabla \psi = \mathbf{i}_\rho \frac{\partial \psi}{\partial \rho} + \mathbf{i}_\phi \frac{1}{\rho} \frac{\partial \psi}{\partial \phi} + \mathbf{k} \frac{\partial \psi}{\partial z}$$

$$\nabla^2 \psi = \frac{1}{\rho} \frac{\partial}{\partial \rho} \left(\rho \frac{\partial \psi}{\partial \rho} \right) + \frac{1}{\rho^2} \frac{\partial^2 \psi}{\partial \phi^2} + \frac{\partial^2 \psi}{\partial z^2}$$

$$\nabla \cdot \mathbf{A} = \frac{1}{\rho} \frac{\partial}{\partial \rho} (\rho A_\rho) + \frac{1}{\rho} \frac{\partial A_\phi}{\partial \phi} + \frac{\partial A_z}{\partial z}$$

$$(\nabla \times \mathbf{A})_\rho = \frac{1}{\rho} \frac{\partial A_z}{\partial \phi} - \frac{\partial A_\phi}{\partial z}$$

$$(\nabla \times \mathbf{A})_\phi = \frac{\partial A_\rho}{\partial z} - \frac{\partial A_z}{\partial \rho}$$

$$(\nabla \times \mathbf{A})_z = \frac{1}{\rho} \left[\frac{\partial}{\partial \rho} (\rho A_\phi) - \frac{\partial A_\rho}{\partial \phi} \right]$$